Human Remains

HUMAN REMAINS

DISSECTION AND ITS HISTORIES

HELEN MACDONALD

Yale University Press
New Haven and London

First published in 2005 by Melbourne University Press, Australia.

This edition published by Yale University Press, London, 2006

For information about this and other Yale University Press publications, please contact:
   U.S. Office: sales.press@yale.edu   www.yalebooks.com
   Europe Office: sales@yaleup.co.uk   www.yalebooks.co.uk

Printed in Great Britain by St Edmundsbury Press Ltd, Bury St Edmunds

Library of Congress Control Number 2006920595

ISBN 0-300-11699-3

A catalogue record for this book is available from the British Library.

10 9 8 7 6 5 4 3 2 1

For Patrick and Anna

CONTENTS

LIST OF ILLUSTRATIONS

ACKNOWLEDGEMENTS

I LEARNED TO BE AN HISTORIAN IN THE DEPARTMENT OF HISTORY at the University of Melbourne. This book has benefitted greatly from the teaching and encouragement I received there, and also from generous advice offered by, and conversations I have had with, others who have engaged with its subject matter in its various guises, in Australia, Britain and the US. Thank you especially to David Goodman, Stuart Macintyre, Greg Dening, Donna Merwick, John Harley Warner, Anna MacDonald, Patrick MacDonald, Joy Damousi, Mandy McKenzie and Ian Duffield. I have also enjoyed many conversations about matters historical (and sometimes irreverent) with Julie Evans, Cheryl Day and Carol Freeman.

Ann Brothers, the Curator of Melbourne University's Medical History Museum, shared her knowledge and aspects of the Museum's collection with me. Chris Briggs, from the Department of Anatomy and Cell Biology, also at the University of Melbourne, talked to me about the process of death in a language I could understand.

Turning words into books is the work of publishers and editors as well as writers. I am grateful to Melbourne University Publishing for their belief in this writing, especially to Monica Dux, who has been a constant, witty and enthusiastic colleague, and to Catherine Hammond for her skilful editing. At Yale University Press special thanks to Robert Baldock for his astute comments and for making *Human Remains* available to its British and American audiences.

I am delighted that the book is published in Britain and the United States, by Yale University Press, for human dissection was a practice that linked the old world and the new, and British-trained medical men made creative use of the new opportunities they found in the Antipodes, dissecting transported convicts, immigrant and indigenous paupers and Aboriginal people. In so doing, they created meaningful personal and professional identities for themselves.

This edition is sub-titled 'dissection and its histories', for historians are exploring how dissection went to work in different historical settings. For the nineteenth century, Ruth Richardson's *Death, Dissection and the Destitute* remains the classic British text. It has now been joined by Michael Sappol's cultural history of anatomy in the United States (*A Traffic of Dead Bodies*), as well as *Human Remains*. Each of these books reveals the historical roots of behaviour and attitudes that, recent scandals reveal, are also present in the twenty-first century. In this way these histories contribute to urgent ongoing debates about the use and abuse of human remains.

I am grateful for the grants that have materially contributed to this book: to the Australian Federation of University Women—South Australia Inc. Trust Fund for the Thenie Baddams Bursary that made a lengthy research visit to Britain possible, to the University of Melbourne for a publishing grant, and to the Department of History, University of Melbourne, for the Lloyd Robson Memorial Award which funded one of my many visits to Tasmania. During these research trips and others, I have met many archivists and librarians without whose expert assistance this book would be the poorer. In Britain, I am indebted to Simon Chaplin at the Hunterian Museum, Royal College of Surgeons of England and Tina Craig, Deputy Head of the College Library; Beverley Emery at the Royal Anthropologial Institute of Great Britain and Ireland; librarians at the British

Scottish Records Office in Edinburgh; and Katya Robinson at the Wellcome Trust Medical Photographic Library.

Over the past five years I have spent a lot of time in Tasmania, so much that Hobart now feels like my second home. Thank you to Margaret Glover, Kim Pearce and others at the Archives Office of Tasmania; Tony Marshall and his colleagues at the State Library of Tasmania (particularly the Crowther Library and the Allport Library and Museum of Fine Arts); librarians at the University of Tasmania; and Jacqui Ward at the Tasmanian Museum and Art Gallery. I have appreciated conversations with Caroline von Oppeln, the late Geoffrey Stilwell, Nicola Goc, Alison Alexander and Stefan Petrow, as well as the members of the Tasmanian Historical Research Association who braved a cold night to listen to me speak about Mary McLauchlan's death and dissection.

Elsewhere in Australia, I thank those who work with the archives and manuscripts at the National Library of Australia, the State Library of New South Wales (Mitchell and Dixson Libraries), and the State Library of Victoria.

Considerable effort has gone into tracing and obtaining permissions for reproducing the images in this book, which are acknowledged in the captions. For any errors or omissions, I apologise and welcome these being brought to my attention.

Finally, thank you to the anonymous referees who gave helpful comments on my several articles based on aspects of this book, published as 'Legal Bodies: Dissecting Murderers at the Royal College of Surgeons, London, 1800–1832', *Traffic: An Interdisciplinary Postgraduate Journal*, No. 2, 2003; 'Mary McLauchlan, Hobart Town, 1830: A Dissection in Reverse', *Lilith: A Feminist History Journal*, No. 13, 2004; 'The Bone Collectors', *new literatures review*, Vol. 42, 2004, pp.45–56; and 'Reading "the Foreign Skull": An Episode in Nineteenth-Century Colonial Human Dissection', *Australian Historical Studies*, Vol. 37, No. 125, 2005, pp.81–96.

A NOTE ON NAMES

As is common in nineteenth-century documents, the spelling of people's names varied. Some of the people who appear in this book—Catherine Welch, Mary McLauchlan, William Lanney and Truganini—have been named in different ways elsewhere. For consistency, they appear here in this form throughout.

INTRODUCTION
PERFORMING ANATOMY

H ERE WE SIT, IN THE BOILER ROOM OF AN ABANDONED
brewery in London's East End. Jack the Ripper territory. It is a
cold November evening in 2002, and we are waiting in a state of
some excitement to view an unlawful piece of performance art. At
last Dr Gunther von Hagens wheels in a steel trolley bearing a large parcel wrapped
in white polythene. The doctor is dressed somewhat oddly in a blue surgical gown
(the mark of science) and he wears a perky black fedora on his head. The hat is
both an artistic quirk and a signature.

Von Hagens is about to perform the first public human dissection in London
since 1832. Standing beneath a print of Rembrandt's *Anatomy of Dr Nicolaes Tulp*
(1632), the anatomist declares that he is returning anatomy to the democratic
science it once was. He unwraps the plastic that swathes Peter Meiss, takes up a
scalpel and begins to cut.

Over the next three hours, our fascinated eyes are riveted on von Hagens's
hands and instruments. With the assistance of knives, a hacksaw and what seems
to be a soup ladle (for removing bodily fluids), he gradually reduces a man to a

shell. At one point the knife slices through the bladder, releasing a stream of sour urine. Some spectators gag. We are not used to this confronting proximity with the dead and what is made of them on a dissecting table. When the performance ends, however, there is almost a sense of anticlimax. Meiss's organs are thrust back into his body, which is roughly sewn together by an underling. It has served its purpose as a piece of event anatomy.

What are we to make of this controversial public performance on a dead human being, which slipped so deftly between science and art? 'I stand here for democracy', proclaimed Gunther von Hagens, linking himself to earlier generations of anatomists who performed such things for an audience like this one.[1] His words made us feel it was our right to watch him carve up Peter Meiss. He did not mention that in Britain those earlier dissections to which he referred were carried out as a form of punishment on the bodies of murderers, which were the only ones made legally available to medical men until 1832. In performing those dissections, surgeons were acting as secondary executioners of the law.

Von Hagens is experienced in justifying his practices (not that there is anything new in that, where human dissection is concerned). In the main he insists that his work is a matter of science, but he cannot refrain from speaking, too, about the artistry it involves. He has used words such as 'edutainment' and told of being approached by West End theatres to host future performances. Not *just* science then, this work on the dead.

No wonder British medical authorities wish this anatomist would disappear. He set off a chain of events that focused attention on how their science was carried out. Journalists consulted files and learned a little about those earlier public dissections which, for von Hagens, were little more than a glib historical reference point. Newspaper stories began to appear about two early nineteenth-century men, William Burke and William Hare, who are notoriously credited with causing anatomy to be regulated in the first place in 1832, soon after it was discovered they had been murdering people to sell their bodies to an anatomist to dissect. (It is much better to relate your work to Rembrandt's *Dr Tulp* than to them.)

In the nineteenth century and earlier, when anatomical 'subjects' were hard to come by, a black market existed in human bodies. Graves were robbed, and surgeons also took possession of people who had died in hospitals when no relatives came quickly to claim them for burial. Now those reading about von Hagens's performance are reminded that anatomy has a disreputable past. Words like 'body-snatching' have re-entered the public realm.

More worrying still to medical authorities, von Hagens's performance came at a time when there were already heightened public sensibilities about medical scientists' activities with the dead. Their science had been tarnished by the discovery that behind the closed doors of hospitals, medical institutions and morgues, people's bodies were routinely harvested for useful parts. There was talk of plastic buckets full of children's organs, stockpiled in what looked like a Liverpool warehouse. Things seemed to be getting seriously out of hand, and von Hagens's highly publicised activities did not help matters at all.

By situating the subjects of dissections at the centre of the narrative, this book differs significantly from most studies about dissection and histories of medicine. Those histories that chart the progressive rise of medical knowledge usually completely dispense with the presence of the dead, upon whom much of that knowledge was made. The deceased are turned into anonyms by instruments and by words, on the dissecting table and in histories of medicine. Yet medical men have always understood the bodies upon which they go to work to be ambiguous material: recently subjects, now objects, called 'subjects' for dissection.

This is because human remains matter. Every society has conventions for dealing with them in a way that involves regulating who has access to bodies and care in their disposal. And when medical men work on the dead, they do so in this knowledge. The human body, whole or in parts, is never just an object like any other, even in a room in which it will be dismembered. It slips between subject and object. We have difficulty in pinning down the moment at which we should begin to refer to such material as 'it', rather than 'she' or 'he'. This uncertainty is especially apparent when a body has died recently, when it is still *some*body.[2]

It is for this reason that during the nineteenth century, surgeon-anatomists tried not to publicly disclose what they did in dissecting rooms. When news of their activities did escape and become public knowledge, they sought to contain the resulting furore by speaking about dissecting in ways that would make it more socially acceptable, pointing to the many benefits that accrued to humankind through their work.

Despite this, many people continued to view dissecting from a critical stance. They understood it was an activity most likely to be carried out on the bodies of people who had been made vulnerable to such use, by one means or another—through poverty, always, but also through the workings of the law and that particular conjunction of interests and knowledge called physical anthropology, in which medical men were supreme.

In von Hagens's public performance, the subject on the table was not a criminal. Nor was he an unclaimed hospital patient, or a man resurrected from his grave. The audience was assured that Meiss had willed his body to von Hagens, though for what exact purpose is unclear. Perhaps he thought he would be preserved in one piece and displayed, like the hyper-real bodies von Hagens exhibits around the world, for which he is justifiably famous. In anatomising Meiss instead, von Hagens broke British law, for such things are only meant to be done under licences issued by Her Majesty's Inspector of Anatomy, and in suitably scientific premises. An old brewery hardly qualifies.

The German anatomist had read the British Anatomy Act. Knowing he was unlicensed to perform a dissection, he chose his words carefully in naming what he would do to Peter Meiss. He advertised the event as an 'autopsy' (though the words 'last performed 1830' revealed that he was speaking about the last public *dissections* performed on murderers in London). The autopsy, von Hagens said, would be carried out 'purely to ascertain the cause of death' and 'find any abnormalities'.[3] That is what autopsies are meant to do. They are not covered by the Anatomy Act and people were not meant to consider that the cause of Meiss's death would have been known six months earlier, when he died, for otherwise no death certificate could have been issued. 'Autopsy', then, was a convenient fiction.

Words have often been used to disguise activities on human remains. Chief Inspector Lemon of Scotland Yard informed the press that the Anatomy Act is very difficult to interpret. Though detectives were among the spectators watching von Hagens work, they did not arrest him. Perhaps it had been decided that the anatomist might welcome such attention.

The organs mentioned earlier in the warehouse in Liverpool were on people's minds. They had been removed without consent from the bodies of children who died in the Alder Hey and other hospitals in that city during the late 1980s and the 1990s. As will become evident when we look at more cases that can only be described as scandalous, many medical personnel have gradually developed an unspoken consensus that they have a right to use the dead for what they consider to be the benefit of the living. It seems the unlawful use of human remains has become a routine matter in medical institutions, despite relevant legislation.

At Manchester University, in the 1980s and 1990s, for example, researchers obtained many brains for a project, without relatives' consent. They defended their actions by insisting such activities were universal practice.[4] As, perhaps, they were. In 2001 in Sydney, Australia, it became public knowledge that organs and other body parts were being removed from bodies lying in the Glebe Morgue and that people were unknowingly burying their relatives incomplete. In addition, at that morgue human remains were unlawfully subjected to brutal forensic experiments involving hammers and knives, activities that had nothing at all to do with performing post-mortem examinations to determine the cause of death.

The parents of the Alder Hey children had asked the British Prime Minister to ban von Hagens's display of human remains—which he had plastinated by replacing their fluids with polymer and then, before the chemical hardened, posing the bodies in positions many commentators find bizarre. Von Hagens fed the controversy when he invited the Alder Hey parents to visit his Body Worlds display and offered to show them around personally. People wondered whether or not this anatomist had gained full consent for his post-mortem use of all those bodies and body parts. At the time, in talking of the forms signed by his donors, von Hagens distanced himself from the doctors who took the children's organs without consent.

But the issue of consent is something von Hagens increasingly needs to emphasise, for others have begun to question how he obtains so many bodies. They note that the plants at which they are processed are located in remote places thought to lack ethical credentials where human remains are concerned. One is in

China, where von Hagens now lives. Another, a medical institute in Novosibirsk, Russia, ships him suspiciously large numbers of bodies, from people who were mentally ill, destitute, or had been imprisoned. This is precisely the kind of human material that has always been most readily available for post-mortem use.

However, the British public was assured that everything was entirely legitimate where Peter Meiss was concerned. And if those assurances were not sufficient to allay suspicions, von Hagens also informed the press that members of Meiss's family—albeit estranged—had also given their permission. It may have been so, but would Peter Meiss have anticipated everything that went on during this public performance over his remains?

Once the word 'autopsy' had been used to justify it, all his human frailties were exposed to four million television viewers. Emphasising the cause of his death meant making public his failed business ventures and subsequent recourse to alcohol. On the dissecting table, he became someone almost deserving of public dismemberment—a lesson for the audience in the error of its ways. He became Peter Meiss, the 'former chain-smoking alcoholic' and 'failed businessman'.[5] That's it, the sum total of a man's seventy-two years. The words objectify him quite as effectively as do von Hagens's hands.

Those who went to watch the performance were conscious of being judged by many others as voyeurs. Some may have taken comfort in the thought that Meiss left his body to von Hagens. Just as well, though, that he was a man and not a woman, for had that been the case, the word 'voyeur' would have taken on other connotations. In fact, von Hagens had been ready to perform on the corpse of a young woman who had died unexpectedly, but there were protests and it all became too difficult. Some kinds of publicity are best avoided.

The anatomist is conscious of how his work on women's bodies may be viewed. He has been criticised for arranging the body of a pregnant woman in a sugges-tive, Venus-like position, though it is likely her pose is just another historical ref-erence point. In August 2002 there was another kind of publicity when a naked living woman suddenly appeared in the gallery among the preserved bodies and mounted a flayed horse. Von Hagens's *Body Worlds Magazine* placed her as protest-ing that there were too few female bodies on display. Fortuitously, a camera was pro-duced to record her demonstration for a larger audience. A photograph appeared in the magazine with the caption 'Lady Cadaver'. In response, von Hagens apologised for this lack of women's bodies among his exhibits, and promised to rectify it. He explained that he had not wished to be accused of 'exposing the female body to male voyeurism'. Soon a 'full body female plastinate with genitalia' appeared among his 'Euro-mummies'.[6]

In these staged performances with human remains, where does the boundary lie between science and art? That is the crux of the matter for many commentators, including concerned medical authorities who understand how difficult it has been historically to persuade the public it is in the interests of science alone that human bodies are dissected. But listen to the spectators at that November autopsy, which many attended for non-scientific reasons—some for the sheer performance of the thing. After all, it was not so different from other shocking forms of British contemporary art. Indeed, the initial sight of Peter Meiss's pale body, which had been preserved in formaldehyde, disappointed those familiar with the technicolour preserved bodies at the gallery next door.

One audience member said he had expected something fresh, and he likened the autopsy to 'going to Christmas dinner and just getting leftovers'. There were complaints that von Hagens seemed 'better at self-promotion than performance', and he was accused of having less charisma than the corpse. 'A little more sensationalism', one man grumbled, 'would have been welcome'.[7] There were enjoyable moments, though. When von Hagens pulled at the sternum with both hands and plunged into Peter Meiss's thorax to lift out his heart and lungs, people cheered.

Those who had come to witness science rather than art also experienced disappointments. Von Hagens, they charged, appeared to be out of his depth. They noticed when he struggled to saw open the skull, and seemed unable to locate the pancreas. A medical student who attended said she felt embarrassed for the anatomist.[8] Those speaking on behalf of British medical institutions also disapproved of the science. One said it was nothing more than a 'grotesque pastiche of a dark but necessary side of the healer's art'.[9] Another called it simply a spectacle, a show. These commentators sought to place some distance between their own, authentic work on the dead and von Hagens's, declaring they had concerns about the respect given to the person who had died—as if the word 'respect' always sat comfortably with such activities on the dead under normal circumstances. The

scientists also worried about why so many spectators wished to witness this ana-
tomisation. They thought it pandered to a morbid voyeurism of the worst kind.

These attempts to distinguish science from art, and real from *ersatz* anatomy,
are telling. Science and art have arguably never been as disconnected as we have
come to assume. Artists have drawn, and still do, upon anatomical science, as when
William Hogarth, in the eighteenth century, attended the dissection of a pregnant
woman and was 'amazingly pleased'.[10] And science is performed in artful ways:
anatomy as ceremony, ritual, exemplary punishment, and staged museum displays.

In drawing a line between science and art, the rich history of these performances is lost. The past becomes instead little more than a promotional tool in
the proclamation that von Hagens is performing the first public autopsy in 170
years. His staff even bring Jack the Ripper into the story. Journalists, too, make
shallow use of history when writing that this event was a 'gruesome spectacle remi-
niscent of the pre-Victorian past'.[11]

Science and art performed on human remains turn subjects into objects,
though never objects quite like any others. Dead human beings retain a certain
unsettling power for a while in the transactions that take place over them. All such
activities with human remains have been part of medicine's shady past, though
—as with all stories—things were rather more complicated than a cursory glance
might reveal.

Cutting into and dismembering human beings is a particularly confronting kind
of work. Those who dissect the dead learn to distance themselves from their subjects to reduce the unease dissecting can provoke. Medical sociologists have written
about how students are inducted into this science. Some students delight in sharing
with outsiders the 'cadaver stories' that are a part of the culture of dissecting; others
take comfort in the belief the person lying before them intended their body to be
used for practice and learning. Some cover certain parts of the body—the face, the
genitalia—or learn in other ways not to see. The face is a particular source of
anxiety. 'I stared at the eyes', one student told Douglas Reifler, 'and forgot to cut
... I was lost'. 'The eyes', another explained to Maureen Capozzoli, 'always look
back at you, even in death'. Cutting the face felt like 'ripping somebody's "self"'.[12]

Some students feel their cadavers have the ability to 'crack' any defensive barrier
the students try to erect between themselves and those who lie on the dissecting

table. But this perception recedes as the skin is breached and the body becomes less recognisably human. Then it becomes safe to see. The cadaver's capacity to disrupt the proceedings fades, until the body being dissected is reduced to something 'not really … human' at all, 'some kind of lesser thing'.[13]

Most of those histories of medicine that discuss dissection focus on the surgeons who practised anatomy 'hands on' in this way to learn their craft. There are, indeed, compelling reasons to dissect the dead. However, the focus of this book is different. It explores dissecting as a cultural activity, rather than the foundational science of medicine, to reveal something about the societies in which such uses of the human body have been made.

From von Hagens's dramatic twenty-first century dissection, we shall move back in time to nineteenth-century England and Tasmania (then known as Van Diemen's Land). These two places were closely related in the British colonial world, and exploring past uses of human remains in them enriches our understanding of twenty-first century dealings with the dead.

Examination of the records reveals performative moments in dissecting that are extraneous to the learning and practice of anatomical science. These show how nineteenth-century medical men came to feel a proprietorial attitude towards human remains, which has continuing relevance in understanding modern-day unethical practices with the dead, of which we often read in our daily newspapers.

In this book, human dissection is narrated through actual nineteenth-century episodes involving the particulars of individual lives. The people who were turned into 'things for the surgeons', and the surgeons themselves, are placed at the centre of this history.[14] Its narrative complements other cultural histories of medicine, and in particular human dissection, which have revealed how particular groups of people (those convicted of crimes, the poor, women and 'native' peoples) became medicine's subject matter. This book builds upon that work, and for the first time the far-flung British colony of Van Diemen's Land is incorporated into the story to reveal how human dissection, learned in Britain, went to work in a different setting. It was one in which British medical men faced a new pattern of opportunities and constraints in practising their craft.

When men trained in this attitude arrived to settle in the Australian colonies, they brought with them the British practices with which they were familiar, and set about the business of making what historian John Warner has called 'satisfying' professional identities for themselves, albeit in vastly different circumstances.[15] In the British outposts of New South Wales (from 1788), then Van Diemen's Land (from 1803), they performed their science in creative ways, seeking advantage in these penal colonies and a system of medicine that was, in the main, government controlled.

Two demographic facts made these colonies advantageous places in which to practise dissecting. The first was that many exiled people died far from families who could otherwise remove them from the surgeons' hands for burial; the second was that they had opportunities to dissect the Aboriginal bodies that were of great interest to Europe's comparative anatomists. This was especially the case in Van Diemen's Land, for that colony's Aboriginal people were thought to be distinctively different from those on the Australian mainland, and by mid-century were understood to be on the brink of extinction. It had the effect of turning their bodies into rare collectables.

This history, then, explores the ways in which certain people's bodies were turned into surgeons' things in nineteenth-century dissecting rooms, were resurrected from graves, and harvested for desirable parts. They were exchanged for favours, posed as pieces of art, and displayed in museums. All of which was as much a social as it was a scientific matter.

When news of these activities with the dead leaked out of the rooms that were meant to contain it, the resulting scandal was captured in letters, diaries, reports of meetings, records of enquiries and in other places. These words are also human remains and they, too, can be explored. In the episodes between these pages, those words reveal the links between medicine, crime and its punishment and the making of racial science. They expose some of the ways in which medicine's great cultural authority was being forged at this time, though as will become apparent, scandals sometimes have unexpected outcomes.

So this is where the following pages take us—into close encounters between nineteenth-century medical men and the human remains which became caught up in the daily jostlings for status and position that were also a part of medicine's past and, current abuses reveal, continue to be part of its present.

CHAPTER ONE
COMPANIONS WITH THE DEAD

IN NINETEENTH-CENTURY LONDON, THOSE WHO WERE HANGED for murder became the property of that city's Royal College of Surgeons, and these were the dissections to which von Hagens referred, in the course of defending his own activities. However, in any given year, very few bodies were made legally available to medical men. In 1831, for example, the number amounted to a mere eleven, at a time when nine hundred men were studying anatomy in the city. So most surgeons found bodies upon which to work outside the law. And they increasingly needed to rationalise their work on human remains.

In 1829, when London surgeon John Abernethy stood to deliver a lecture to his students at St Bartholomew's Hospital, he began by informing them '[t]here is but one way to obtain knowledge ... we must be companions with the dead'.[1] Many believed that only through hands-on dissecting could 'views become essentially our own', and 'we dare give them the name of knowledge'.[2] Direct observation of the dead human body provided alert men with valuable opportunities to trace the effects of disease, and this complemented their clinical practice. One of London's best surgeons at the time, Sir Astley Cooper, articulated it in the following way:

by examining Dead bodies, we become acquainted with the changes produced by Disease, with its nature whether curable, or incurable; if it be of the former description we are enabled to form an opinion respecting its best mode of treatment, and if of the latter, we avoid giving unnecessary torture to future Patients with the same disease.[3]

But learning medicine in this empirical way, through sight and sound and touch, had practical effects. As students noted, it meant that 'every medical man is compelled to ... go to work as if the science existed in its original state of chaos and confusion, in order to fix for himself with exactness and certainty its elementary facts and principles'. Each man needed to find for himself a way 'out of this darkness'.[4]

By the mid-eighteenth century, experience in dissecting was conventional practice for London's surgeons. However, in Britain, in contrast to France (where revolutionary edicts made a plentiful supply of bodies available to empirically minded medical men), creative ways of obtaining human bodies needed to be found. What medical men did with the dead contravened deeply held social beliefs about how human beings should properly be dealt with, post mortem. It mattered that bodies were treated with respect and buried whole for all kinds of reasons, including an expectation that the dead would rise in bodily form on God's final judgment day. There was also widespread anxiety over how precisely to determine the point beyond which a person was certainly dead.

In England the only legal supply of bodies came from the scaffold. Under *An Act for better preventing the horrid Crime of Murder* (25 Geo. 2, c. 37, 1752), or Murder Act, either dissecting or gibbeting became an additional punishment meted out to murderers alone, due to the heinous nature of their crime. From this time the College of Surgeons had the right to the bodies of all convicted murderers executed in London. However, the fact that there were very few in any given year meant that medical men and their students inevitably needed to obtain most of their supplies elsewhere. They went to work on corpses that had been stolen from graves soon after their burial, or that lay unclaimed in hospital dead houses. Much more rarely, they purchased the bodies of people who had been murdered to turn them into objects for sale.

What did the College make of these few bodies that were legally available in London for dissection? Who were these people, and to what uses were they put? Exploring these matters reveals that their bodies were not used for science alone, but were turned into material for experiments, and gifts that bound men working in London's charitable hospitals to the College. Let's spend some time in the house the College hired close by the scaffold, before moving on to unlawful dealings with bodies in hospitals, where patients were turned into subjects for dissections, and private anatomy schools that obtained bodies via a black market in human remains.

DISSECTING MURDERERS

Until 1832, London's College of Surgeons had been receiving all the bodies of those executed for murder in that city since 1752. These public dissections were crafted social events. Astley Cooper, who carried them out between 1793 and 1796, at Surgeons' Hall, found that the College's theatre was constantly crowded, and the applause excessive.[5] Executions, too, were public spectacles that attracted huge and rowdy crowds. They were usually performed at eight o'clock on Monday mornings, following which the body was left dangling at the end of the rope for an hour before being carted to the College's house. It was accompanied on that short journey by both the executioner and London's Sheriff, and was received by the College Master (later President), dressed in his official regalia.

Between 1800 and 1832, William Clift, the Conservator of the College's Hunterian Museum, performed these dissections. Though not a surgeon himself, Clift was exceedingly skilled in dissecting, having been apprenticed to England's premier comparative anatomist, John Hunter, whose collection formed the basis of the museum. In dissecting murderers, Clift worked under the direction of the College Master. Of the forty-five dissections performed during these years, twelve were brief affairs comprising only a 'proper' examination, which meant nothing more than making an incision over the sternum. This was a theatrical cut, made to mark the College's treasured monopoly over these subjects. After that, these bodies were turned into gifts for well-connected surgeons or pupils in London's hospitals and, more rarely, in the city's private schools of anatomy.

Public Execution, 1807, artist unknown. At this execution of
Elizabeth Godfrey, Owen Haggerty and John Holloway, a pie
seller fell over in the crowd, setting off a riot that left many
people injured and some trampled to death
(*Gillen,* Assassination of the Prime Minister).

However, most of the College's dissections were much lengthier affairs which provided opportunities to experiment on the remains of people who had recently died.

Galvanising George Foster, 1803

The social anxieties about the precise boundary between life and death that were common in Europe at this time were expressed in novels and poems, and also circulated in dark tales about people who had been buried alive. Certain people who had been executed had even later revived on the College's own dissecting table.

Some experiments performed at the College aimed to determine the time of death with certainty. They were so violently crude that Clift clearly found his job distressing. In an uncharacteristically stumbling hand, he recorded what he was instructed to do to the body of Martin Hogan as it lay on the dissecting table in

1814, which was to thrust a needle into each eye to see if that produced an effect. Other investigations were undertaken in a more systematic way as College men sought to understand whether an absence of obvious animation was a sign that the life force had merely been suspended or was irretrievably extinguished. And although this was never stated in the context of the College's work on murderers, they were also wondering whether it was in the power of medical men to return people to life.

In 1803 the College invited Professor Giovanni Aldini to perform galvanic experiments on the body of George Foster, who had been found guilty of murdering his wife and child by drowning them in the Paddington Canal. Aldini required access to the bodies of people who had died very recently, in the belief that these still held their 'vital powers'. In contrast, those who had died of disease might have 'humours' which would resist his experiments. Later, writing up his London work, Aldini spoke admiringly of England's 'enlightened' laws, which provided murderers with an opportunity to atone for their crimes by such uses of their bodies after death. He argued that galvanic experiments were especially in the interests of a *British* public, for Britain was a commercial and maritime nation, filled with rivers and canals. When British people drowned, he said, galvanism might provide the necessary 'means of excitement' to return them to life.[6] Dealing with the dead in controversial ways always requires some form of rationalisation. It is part of the process through which access to bodies is socially negotiated.

Reading the records of this scientist's work at the College in 1803, it is not difficult to see why others believed such men liked to play at being God. Always conscious of his audience, Aldini made the dead perform tricks. He boasted that in Europe he had once placed the heads of two decapitated criminals on separate tables, then connected them with an arc of electricity to make them grimace to such an extent that the spectators were actually frightened. He had also made the hand of a headless man clutch a coin and throw it across a room.

The College provided Aldini with an opportunity to undertake some new experiments on George Foster, whose body had been left hanging for an hour in temperatures two degrees below freezing point. When it was delivered to the College's house in West Smithfield, College Master Keate ceded his authority to

Galvanising Human Remains, 1804, by Giovanni Aldini
(*Aldini*, Essai théorique et expérimental).
Courtesy of the Wellcome Library, London.

direct what followed to Aldini, who applied arcs to various parts of the body over the following hours to make George Foster perform. His jaw quivered, his left eye opened, and his face convulsed. When conductors were applied to his ear and rectum, the resulting muscular contractions 'almost [gave] an appearance of re-animation'.[7] One hand clenched and the right auricle of the heart contracted. Aldini's audience was amazed by such signs of animation.

These experiments continued for more than seven hours after the execution. While Aldini denied an intention to reanimate the corpses upon which he went to work, everyone in that room would have considered it a triumph had he managed to do it. *The Times*, whose reporter probably witnessed the day's work, noted that a principle had been discovered 'by which motion can be restored to Dead Bodies'.[8]

Such possibilities had not been in the minds of England's legislators when they worded the Murder Act. Dissection was meant to result in the mutilation of the dead, not their resurrection. Bringing executed murderers back to life might have been a matter of congratulation for the man who achieved it, but it would have been a complicated problem for the law. Aldini himself spoke ambiguously about his intentions. He said the object of his experiments 'was not to produce re-animation, but merely to obtain a practical knowledge how far Galvanism may be employed … to revive persons under similar circumstances'.[9]

Regulating Dissection: John Bellingham, 1812

That surgeons, like executioners, went to work on behalf of the law in punishing murderers was a matter of discomfort to many of them. They thought the association between themselves and the hangman was degrading, and preferred a self-image that tied dissection closely to the promotion of medical science, rather than to inflicting indignities on human bodies on behalf of the Crown.

With this in mind, the College's dissections were supposed to follow a certain dignified path, but the fact that they often did not is revealed by the College's attempt, from 1812, to officially regulate them. As Giovanna Ferrari has noted in her work on Bologna's anatomy theatre, the very existence of such regulations reveals the kind of undesirable behaviour that was actually taking place.[10]

The College's 'Regulations Relating to the Bodies of Murderers' were written on 8 July 1812, at the first meeting of the museum's Board of Curators following the dissection of Britain's only assassin, merchant John Bellingham. Bellingham had murdered Prime Minister Spencer Perceval in the lobby of the House of Commons just a week earlier. The assassination was initially perceived as the beginning of a political revolt against an unpopular government, and it caused mayhem in London's streets. When Bellingham ascended the steps to the scaffold a week later, he was greeted by cheers, and the government felt the need to station 5000 troops nearby.

The College's house was crowded with spectators for Bellingham's dissection. We can catch a glimpse of the kind of behaviour that prevailed that day in the regulations that were enacted soon afterwards. From this time on, they instructed, 'Disorder, and Interuption [sic] during the Dissection' were to be prevented by admitting only people who were accompanied by a Member of the Court of Assistants, and in addition, from now on only the Master or one of the two College Governors could give a 'specific Direction ... respecting the Dissection'.[11]

William Clift opened each cavity of Bellingham's body, and certain matters were recorded. The stomach was found to contain a small quantity of fluid ('which seemed to be Wine'); the bladder was empty and contracted; the penis 'seemed to be in a state of semi-erection'; and the brain was found to be 'firm and sound throughout'. Some audience members may have been surprised by this, expecting to find material signs of madness in the brain of a man who had murdered a prime minister. Bellingham's stomach and left testicle were sent to the College Museum.[12]

The Master also directed that some experiments be undertaken on this body. Still interested in animation and the dead, the College men wished to explore how long a heart could be made to move after the moment of death. This was the beginning of such experiments at these public dissections, and they continued until 1827. Sometimes the men who were present were content to observe a heart pulsating of its own accord. At others, they artificially stimulated it. Some hearts were cut out of bodies, placed in a saucer and watched. When they finally lay still, they were nudged with a scalpel to see if they would begin to move again.

The motion of Bellingham's right auricle was observed with care. Its movement was assessed as being 'not strictly a contractile action, diminishing in any Sensible degree the cavity of the Auricle', but 'undulating and weak, sometimes beginning at the right extremity of the Auricle & moving to the left: at other times com-mencing & proceeding in the contrary direction'.[13] The surgeons experienced one of their greatest triumphs with this man's heart, for its right auricle continued to move 'without the application of any Stimulus, during the period of nearly four Hours from the Time of Execution; and for about an Hour Longer, upon being touched with a Scalpel'.[14]

When the College was finished with Bellingham's body, President Sir William Blizard turned it into a gift for a Mr Stanley, a pupil at nearby St Bartholomew's Hospital. Murderers' bodies were often given to men working or learning in London's charitable hospitals. These gifts were one way in which the College main-tained its close relationships with London's most powerful surgeons.

Bellingham's body seems something of a prize for a pupil, given the fame of this subject and the controversy that surrounded his speedy trial and execution, both of which attracted large audiences. Perhaps whatever else had gone on in the College's dissecting room—the kind of behaviour that had precipitated those new regulations—had resulted in there being so little left of the body that few other medical men would have wanted to receive it.

Seven years later, it had become apparent that the 1812 regulations needed to be strengthened. The new rules stated firmly that all men concerned with the dissecting of murderers were to act in a way that was 'as conducive as possible to general Benefit by promoting the Designs of the College, with reference to the Museum and Theatre, and, otherwise, facilitating the cultivation of anatomical and chirurgical knowledge'. From now on, three College Members at least were to decide to whom murderers' bodies were to be given, and none was to be 'taken or removed from the appointed Place of dissection until the Dissection thereof shall have been duly performed'.[15] But despite the clarity of this Standing Order, in practice individual College men continued to pass the bodies of murderers to favoured individuals elsewhere.

Dissecting Women: Catherine Welch, 1828

It was rare for a woman to hang for murder in England at this time, so of the bodies received at the College between 1800 and 1832, only seven were female.

Five of these left the College in relatively pristine condition, having received only the 'proper' dissection. All but one of the five made valuable gifts, for these women were young and healthy when put to death, and there was much to learn on such subjects for dissection at a time when women were viewed, anatomically speaking, as being defined by their organs of generation. The fact that four of these five women were in their reproductive years counted for something.

At this time, medical men could only learn about the internal anatomy of the female reproductive system inside the bodies of dead women. Yet surgeon-anatomist Charles Bell informed his students that women's bodies were difficult subjects to investigate on the dissecting table.

> To give a full and comprehensive view of the diseases incident to the female pelvis would lead to a very long dissection ... [they] are difficult to be understood; and the explanation of them requires a complete knowledge of a very delicate and minute piece of anatomy, together with much practical dexterity. It is indeed a subject so extensive, that it cannot be fully treated of on the present occasion.[16]

This, in a two-volume work that claimed to explain the anatomy of the *human* body.

There were social as well as scientific reasons for medical uncertainties about the female body. In lectures and dissecting manuals, physicians and surgeons impressed upon their students the need for 'decorum' when working on the bodies of their well-to-do female patients. Decorum dictated that, at least where the genitalia and accessible generative organs of private patients were concerned, a man must carry out his work blind. Astley Cooper instructed that inserting a catheter into a woman's urethra to draw off urine 'ought always to be done under the bed cloathes [*sic*]'.[17] John Haighton informed students that they should examine the bodies of pregnant women by 'feeling', 'without seeing the part'.[18] Students needed to learn the comparative location of urethra and vagina by heart, orientating themselves from the position of the clitoris, lest acts performed blind should go horribly wrong.

No such concerns with decorum were necessary when investigating the body of a woman who was dead. On 14 April 1828, 24-year-old Catherine Welch was carted from the scaffold to the College's house, after being executed for murdering her six-week-old son in tragic circumstances. Deserted by the man who had impregnated her, she married another who, when he found she was pregnant, refused to support her while 'the child existed to publish his disgrace'. According to the reporter for the *Evening Mail*, who had attended the execution, Welch was 'a fine young woman of stout and particularly healthy appearance'. He noted that she had struggled greatly at the end of the rope 'for some minutes', though there are no signs on her face of this slow death by asphyxiation.[19]

The College performed only a proprietorial cut on Welch's body, which was then given to Charles Bell, who ran a private school of anatomy in Windmill Street. Thanks to the sketches William Clift and his son made of those who came to lie on the College's dissecting table, no matter how briefly, we can catch a glimpse of what Bell's students saw as they crowded around the table on that day.

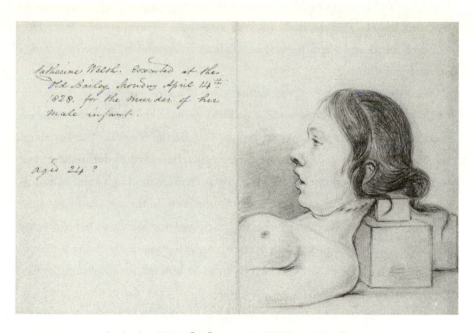

Catherine Welsh [sic], 1828, by William Clift (RCS,
Clift's Heads of Murderers, 1807–1832, Box 67.b.13).
Reproduced by kind permission of the President and Council of
the Royal College of Surgeons of England.

Catherine Welch became a subject for dissection within a few weeks of giving birth to and last suckling her child, so a number of interesting things could be demonstrated on her body. Contemporary dissecting manuals taught students the process. Bell would begin with the surface anatomy. First, the mamma, a gland 'peculiar to the female', of 'a rounded form', which adheres loosely to the surface of the large pectoral muscle.[20] He would instruct the students to draw the nipple out to make the excretory ducts visible to the eye, and notice the way the *areola* varies in colour at different times of life, being of a darker colour during pregnancy than at other times. In his own dissecting manual, Bell directs students to attend particularly to the mamma in women 'giving suck'. Catherine Welch fell into this category, so her breasts would have been larger than common, 'hard and troublesome to the touch', the nipples thick and strong.[21]

The students may even have tried to replicate a test performed by a surgeon, Mr Holmes, who gave evidence against Welch at her trial. When she denied having a child to either suckle or murder, he proved otherwise by extracting milk from her breasts. Perhaps Bell's students attempted to do the same. It could be achieved, instructed Farr's manual on medical jurisprudence, either by 'pressure' or 'suction'.[22]

Now they could move on, past further signs of her recent delivery—a relatively rough and flaccid abdomen—to the external parts of the organs of generation. Separating the labia, they could explore the *Clitoris*, the *Nymphae*, the *Vestibulum*, the *Meatus urinarius*, and the orifice of the vagina. Bell would compare the clitoris to the male penis. Perhaps he sliced it in two for a more thorough examination. To inspect the perineum, the students would fix the body 'in the same position as that for the operation of the stone'.[23] This meant either lying Welch on her back, parting her legs and placing them over her head (or, as she was dead, sawing them off). Bell cut off her breasts, too, and sent them to be injected for display in the College's Museum.

Bell's dissecting manual explains how he would then have opened the body. He advised students to 'hold the knife easily in the hand, not rigidly ... lay the edge fairly to the part ... cut with a steady and uniform stroke ... but the best rule in this, and all such operations, is to avoid all affectation of manner'.[24] He taught his students how to recognise further signs that a woman had recently given birth.

Her vagina would be distended, and there might still be signs of the presence of a different substance from the common menstrual flux. In examining the orifice of the uterus, the students would notice that it was still soft and open, not yet collapsed to its natural shape. The extent to which the uterus was exposed during a dissection depended upon whether or not the subject's bladder was full. In the body of a woman who had been hanged, it would probably have been empty. So the advice of surgeon John Flint South's *Dissector's Manual* would have been followed: distend the bladder and stuff the rectum to make viewing what lay within the pelvis easier.

At this time, knowledge about the ovaries was still a source of interested speculation rather than established fact. They were said to correspond to the testicle in the body of the male. Some believed they carried the rudiments of the foetus; others held that miniature human beings lay in a man's sperm. Working on the ovaries was something that could only be done on the bodies of dead women, and what men learned there could not, before the invention of anaesthetics, assist in dealing with ovarian disease in the living. For Charles Bell, this made deep knowledge about the ovaries unimportant, a kind of useless acquisition.

Slowly, beneath these enquiring minds and hands, Catherine Welch disappeared. In return for the gift of her body, the College expected to receive from Bell a report that would contribute to anatomical and surgical knowledge. He sent none, which was a source of ongoing ill-feeling. Two years later, the College Secretary was still writing to request an official report of this particular dissection.

Dissecting the 'Female Burker': Elizabeth Ross, 1832

In 1828, it was discovered that two Irishmen living in Edinburgh, William Burke and William Hare, had murdered sixteen people in order to sell their bodies to the private anatomy school of Dr Robert Knox. Burke and Hare had developed a system of murder (or 'burking') that left few marks on their victims' bodies, and it is unlikely that Knox suspected he was dissecting victims of murder. Nevertheless, the public furore that followed the discovery spelt the effective end of his career.

At the Royal College of Surgeons in London, William Clift was deeply interested in the phenomenon of burking. This is not surprising, given the publicity these crimes attracted and the additional fact that some similar English cases were discovered in London in 1831. The London burkers, John Head and John Bishop, came to lie on the College's dissecting table after their execution, as did Elizabeth Ross a year later. She was the only female 'Convicted Burkeite'.[25]

As was his habit, Clift sketched all three before the dissections began. If he was tempted to make these people's faces illustrate some kind of physical propensity for their crime, as others did at the time, he resisted it. Clift's sketches contrast, for example, with some made of Burke and Hare, in which one artist, influenced by phrenology and theories of physiognomy, gave Burke's skull the shape of a man in thrall to the animal passions, and turned Hare into a cunning fox. (Hare *was* cunning, for though equally involved in the murders, he gave evidence against Burke and escaped any punishment for these crimes.) Clift portrayed the English burkers in a much more open-ended way, which contrasted with the contemporary word pictures made of them in newspaper accounts, which turned Head, Bishop and Ross into monsters whose crimes were incomprehensible.

Clift's sketches, made quickly before he was instructed to take up the tools of dissection, often problematised contemporary accounts of the murders that brought the people who lay before him to the dissecting table. Elizabeth Ross's conviction was based on very shaky evidence, most of it provided by her twelve-year-old son to save his father, Edward Cook, from execution. The son insisted his mother had acted alone in smothering the family's elderly lodger, then hawking the body to anatomists at London Hospital. Ross denied the crime. She said her son had been coached in what to say at the trial, and that she had left the lodger alive in the company of both him and her husband.

Her refusal to confess to the crime before she was executed left loose ends that the press tried to tie together. They sought the opinions of her neighbours, who remembered noticing that when Ross moved into their neighbourhood, cats began to disappear. They also said Ross was a gin-drinker and a thief. In the absence of the victim's body, which had presumably already been dissected at London Hospital, newspaper reports tried to establish this woman's guilt by relating her crime to the Head and Bishop murders a year before. Despite what one newspaper called its most rigid inquiries, it reported that her only motive must have been a wish to

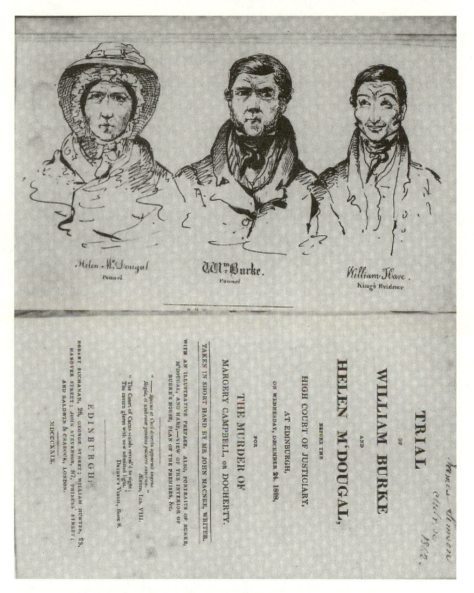

William Burke and William Hare with Helen McDougal,
1828, by D. M'Nee (detail from a pamphlet in Barzun,
Burke and Hare). *Courtesy of the New York Academy of*
Medicine Library.

profit from selling a body, since everyone knew from the earlier trial that money made by such a murder was relatively easy.[26]

The newspaper gave Ross both a nationality (Irish, like Burke and Hare) and a physical appearance to match this understanding of the crime. She was said to be a 'large, raw-boned, and coarse featur'd Irishwoman' of 'masculine proportions and strength'.[27] This picture in words made it believeable that Ross had single-handedly murdered her victim (though it had taken both Burke and Hare to hold down and smother theirs, some of whom struggled mightily), then carried the dead weight of the body through the streets to London Hospital to negotiate a sale.

Clift's pencil sketch of this woman tells a different story to that which appeared in newspaper accounts. There is no sign here of masculine strength. Clift has included in this portrait his subject's upper arm and shoulder. No outline of muscles appears, and the words that accompany the sketch tell us that Ross was

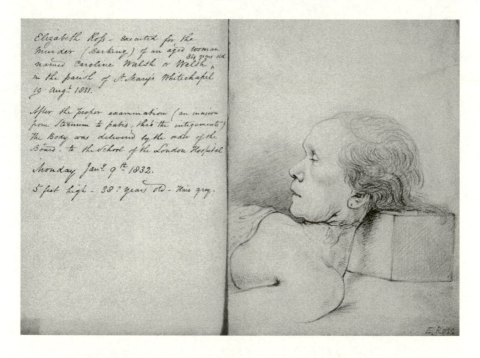

Elizabeth Ross, 1832, by William Clift (RCS, Clift's Heads of Murderers, 1807–1832, Box 67.b.13). Reproduced by kind permission of the President and Council of the Royal College of Surgeons of England.

only five feet high (though she was measured on the dissecting table, and 'long' was a more accurate word). She does not seem to have been a substantial woman. Although her head is thrust forward by the block of wood beneath it, there is hardly a sign of a double chin.

Ross was only thirty-four years of age when this sketch was made, but her sparse hair makes her appear much older, and the rope from which she hanged has left a mark around a wrinkled neck. This woman's life had given her the look of an old woman. Taken together, the physical facts in Clift's portrait suggest that Ross was no murderer of masculine strength. So does that hole in her earlobe, which indicates she cared sufficiently about the way she appeared in the world, as a woman, to decorate her face with earrings.

Ross's breasts are covered by a cloth, which is unusual in Clift's portraits of women. I think it indicates he made this sketch after, rather than before carrying out the 'proper' dissection on this body, for in this case, unique for a woman, the dissection comprised 'an incision from Sternum to pubis, thro' the integuments'.[28] This was the way a man would be cut, if a full dissection were about to follow. A woman was normally opened differently, to leave a triangular flap to 'fall over the parts of generation'.[29]

Clift would have been following the President's instructions: one slashing mark of contempt for a woman whose crime reminded the public that surgeons stood in ill-repute where gaining access to the dead was concerned. Or perhaps Elizabeth Ross, though only thirty-four, was assessed as being too old to make her body suitable for illustrating the reproductive organs.

This, then, was something of the flavour of those last public dissections, to which Gunther von Hagens referred in such a cavalier way in November 2002 when promoting his own performances with the dead. This punitive cutting was a complicated business that rewarded those who performed it in such authoritative ways. Turning human beings into things for the surgeons attracted the admiring gaze of audiences and bound the powerful men attached to the Royal College of Surgeons into reciprocal relationships with anatomists and surgeons working in London's charitable hospitals and extramural, private schools of anatomy.

'A KIND OF NECESSARY INHUMANITY'

In speaking of dissection (in lectures, dissecting manuals, anatomy texts and the pages of the medical press) medical men placed it as a self-sacrificing activity that they performed on humanity's behalf. It was a slow, tedious, disagreeable form of work carried out by men who had to learn to overcome their natural aversion to touching the dead and enter the filth and stench of the dissecting room. William Hunter, one of London's best-known eighteenth-century physician-surgeons, informed his students that anatomy learned through dissecting would familiarise the heart 'to a kind of necessary Inhumanity'.[30]

Anatomy was medicine's essential science, and if it were not learned on the dead, then knowledge would need to be obtained by 'mangling the living'.[31] Without anaesthetics to render a patient unconscious during an operation, a surgeon's most useful skills were speed and dexterity, and these could only be obtained by constant practice on the dead, or by vivisecting animals. Deeper operations were only performed in extreme emergencies, in the knowledge that the patient would probably die. Astley Cooper said he personally 'would not remain in a room with a man who attempted to perform an operation in surgery who was unacquainted with anatomy'. He advised the members of a select committee established to examine such matters, that the regulation of anatomy and the provision of an increased legal supply of subjects for dissection was 'not our cause, but your's'.[32] The editor of the *London Medical Gazette* agreed. Medical men's motives in seeking access to the dead were simply 'a disinterested regard to the general weal'.[33]

In this debate, surgeons and anatomists necessarily contrasted their need for bodies to dissect in the cause of science against many people's horror of dissecting. Thomas Southwood Smith thought the 'uninstructed and ignorant' were 'full of prejudice and error', their feelings about the dead requiring 'control, and sometimes even sacrifice' in a nobler cause.[34] When suggesting exactly whose bodies should be sacrificed in this way, the *London Medical Gazette* favoured using the remains of 'those unfortunate persons who have none to own them in life or to claim them when dead'. It was, the *Gazette* charged, 'unreasonable and absurd' to object to this solution, for dissection caused no 'injury to the deceased', whose identity was not 'hazarded by the dissection of his remains'. The article declared:

> [T]o suppose that the immaterial spirit still clings to the mouldering body … is a notion calculated for the times of ignorance and darkness … No—if the dead can, indeed, be disturbed by any thing inflicted on the body, it is the rats that gnaw it, and the unmannerly sexton, who knocks it about with spade and shovel, that are the real delinquents, and not the man of science, who takes it from these disturbers to make it subservient at once to humanity, knowledge, and religion.[35]

In any event, in promoting the use of bodies unclaimed by relatives when they died, the *Gazette* was speaking of a practice already firmly established. Surgeons attached to England's charitable hospitals had opportunities to dissect that were the envy of others. During the late eighteenth century, these hospitals, which serviced the poor, had gradually become places where pupils learned to be medical men. Holding the prestigious position of an 'honorary medical officer' in one of them brought many rewards, though these were not paid positions. By donating their time to the treatment of destitute patients, these men and their private, fee-paying pupils were given opportunities for clinical practice on the bodies of patients, alive and dead.

Clinical medicine looked forward to dissection, which enabled medical men to understand the cause of a patient's illness. And post-mortem examinations carried out to establish the cause of death had a way of becoming a full dissection in teaching hospitals. According to medical historian Charles Newman, it became 'a race whether [relatives] or the pathologists got there first'.[36] In Bristol during the eighteenth century, for example, it is known that poor patients whose bodies were not claimed within just a few hours of their deaths were almost always taken from ward to dissecting room. A contemporary student notebook (1822) shows that the only patients who died and were *not* dissected in that city were those with vigilant friends who sat with them until they were taken out of the institution in a coffin.[37]

Even though these shady dealings provided medical men and their students with large numbers of bodies for their practice, there was never a sufficient supply. Many human remains were purchased from a group of men whom surgeon-anatomists both needed and despised. Britain's 'resurrection men' dug up bodies from their graves and sold them to anatomists. Such purchases were, according to the presiding judge at William Burke's trial, 'the invariable and the necessary practice of the profession'.[38]

Some distance was placed between the resurrectionists and those whose schools they supplied through the anatomist's porter and his students. It was the porter's job to negotiate prices and receive the bags and chests containing bodies. We catch a glimpse of the range of a porter's tasks from a list made by Robert Knox's students in 1828. David Paterson's duties at Knox's extramural school included:

> keeping the door, cleaning and sweeping out the rooms, putting on and mending the fires, scrubbing the tables, and carrying away and burying the offals of the dissecting rooms; washing and cleaning subjects preparatory to their being brought into the class-room; attending on the students, and doing little jobs for them—such as cleaning and scraping bones, getting their dissecting clothes washed, which was done by his mother and sister; he had likewise to go messages, and be ready at all times to receive packages [i.e. bodies] and go for them.[39]

It is useful to notice how, in speaking of the porter's dealings with the dead, the students erase their own. Many of them considered it a prank to rob graves, exciting sport. Dealing with the resurrectionists was also a part of their educational experience.

To many people in Britain at this time, the medical men who purchased bodies for dissection seemed to be monsters, for they were acting far outside what was acceptable to the society in which they lived. Their dealings with the dead—purchasing, grave-robbing and the like—caused public outrage and rioting against anatomy schools. Historians and nineteenth-century commentators speak of medical men and students as being a brutish group. Charles Newman says they were 'to put it mildly, appalling': foul-mouthed, indecent, callous, cynical and physically dirty.[40] Contemporary caricatures portrayed them this way. Surgeons were also pilloried as men who sought the right to 'bleed, cut, draw, lance, probe, saw, hack, mangle, tear, blister, burn, embrocate, fumigate, mend-a-pate, potion, lotion, lotion-palm, pocket, charge and kill'.[41]

*The Persevering Surgeon, date unknown, by Thomas Rowlandson.
In this dissecting room a leering surgeon hovers over the naked
body of a young woman, one claw-like hand reaching for her
breast. The image reflects popular fears of what might occur in
dissecting rooms. Reproduced by kind permission of the President
and Council of the Royal College of Surgeons of England.*

Students became infamous for the kinds of pranks that made many wonder about their humanity, as when a London pupil climbed onto the roof of a house and dropped an amputated leg down its chimney, which resulted in a riot. Sir Squire Sprigge, in his biography of early nineteenth-century medical reformer Thomas Wakely, saw the dissecting room as being central to the whole 'tone' of the medical school fraternity, at a time when anatomy was the most important part of a surgeon's education. The dissecting room, he said, was the common meeting place for students, 'like the studio in the life of the art student'. In such places, a kind of camaraderie was established and maintained, and 'every common interest

The Dissecting Room, date unknown, by Thomas Rowlandson.
This is thought to be a representation of William Hunter's
dissecting room. It is filled with skeletons, the busts of famous men
and a mass of bodies, dead and alive. Corpses are draped on tables.
A head carelessly lolls over the end of one, mouth gaping while
students open the abdomen. Five or six men surround each corpse.
Another body lies in a crate on the floor, where it is being prepared
for their use. Reproduced by kind permission of the President and
Council of the Royal College of Surgeons of England.

[was] discussed'. Students developed the kind of familiarity with 'repulsive objects' that destroyed a proper regard for the common decencies of life, especially a respect for mortality. Going to work in the dissecting room was a rite of passage in which it became 'almost necessary' for each man to assume a 'regular and accepted tone of low shiftlessness', and it was 'an act of *esprit de corps* to transmit [this] to his juniors'.[42]

Some early nineteenth-century medical students have also left records of the distaste they felt at the actions of their colleagues. Henry Acland, a student at St George's Hospital in London in 1830, thought his fellow students were men of low habits. William Dale wrote in 1840 that 'drinking smoking and brawling were the very *rational* occupations of the dissecting room … [and] it was no uncommon thing to see a regular battle among the students, parts of the human body forming their weapons'.[43]

Sex in the Dissecting Room: Mary Paterson, 1828

At St George's Hospital, medical student Robert Christison was shocked by the disrespect with which bodies were treated on the dissecting table, especially those of women. Dissecting, as historian Alison Bashford has noted, was a cross-gendered activity: men dissected women.[44] Christison wrote of the popular 'Dr B—', who lectured on obstetrics in circumstances of 'indecency without any qualifying wit'. He contrasted Dr B's lectures with those delivered by John Abernethy ('pure thoughts, sound reasoning, beautiful language and noble delivery'). Another student, James Paget, was horrified by his fellow students at St Bartholomew's and said his teachers told 'utterly indecent and dirty stories'.[45] The dissecting room was a place in which gross indecencies were performed and not only by medical men. A porter at St Bartholomew's was said to have raped the corpse of an attractive fifteen-year-old girl in front of medical students.[46]

Artists also had entrée to Britain's nineteenth-century dissecting rooms, where they learned to perfect their representations of the human form. William Hunter, the first anatomist who held the position of Professor of Anatomy at London's Royal Academy of Art (in 1768), said it was only by possessing a thorough knowledge of anatomy that an artist could create such an 'exquisite pitch of delusion that the Spectator should mistake it and think it is what it only represents'.[47] Others decried such artistic immersion in anatomy, believing that paying too close attention to the insides of the human body resulted in an inadequate representation of external form.

This was a subject to which Edinburgh anatomist Dr Robert Knox had given considerable thought. In his *Great Artists and Great Anatomists* (1852), he wrote of the link between dissection and destruction, which was something others promoting the usefulness of dissection for artists refused to acknowledge. Against utilitarians, he insisted that the process of dissecting was a hideous one, in which beauty quickly disappeared from view. He worried that any artist who paid close attention to what he called the 'mangled and dissected dead' would be unable to adequately represent the living. No artist, Knox said, should be sent 'into a charnel-house, called a dissecting room', for in such a place 'his mind may become accustomed, by frequent contemplation, to all that is detestable; familiar with horrors, with the emblems of destruction and death'.[48]

And Knox was a man who would know, for he spent seven hours each day dissecting. It was a process that had disenchanted him about the inside of the human body. He thought it was all hideous shapes. For him, beauty lay in external form alone and '[w]oman', he wrote, 'presents the perfection of that *form*, and, therefore, alone constitutes' the absolute 'perfection of Nature's works'.[49]

Perhaps this is why, when William Burke brought him the body of eighteen-year-old Mary Paterson in 1828, this anatomist could not bear to cut her open or turn her over to his students to dissect. Instead, he invited an artist to sketch her and then preserved her whole in a tub of whisky for three months. Only then did she become one of the mangled and dissected dead in his school.

Knox would never shake off the effects of the scandalous discovery that he had purchased the bodies of sixteen people who had been murdered, though he had not directly received their bodies himself, and there is no evidence he knew of how they had died. Mary Paterson was Burke and Hare's third victim. They murdered her in their characteristic style, offering her whisky and then, when she was unconscious from its effects, one held his hand over her nose and mouth while the other lay on top of her to deter any struggling while she smothered. Next they packed her body into a tea-chest and carried it through the streets of Edinburgh to Knox's school in Surgeons' Square, where the porter, David Paterson, gave them £10 in exchange for this subject for dissection.

If we were to believe, as has so often been asserted by medical men, that the activities in dissecting rooms were only about making medical knowledge in a self-sacrificing way, what sense could we make of a sketch like the one below? It is a post-mortem portrait of Mary Paterson, a young woman who had made her living on the streets of Edinburgh by selling the use of her body for sex to those who could afford to pay for it—as, it turns out, may at least one of Knox's students, William Fergusson, who was later to become Queen Victoria's Surgeon-General.

Some of these students, as well as David Paterson, retrospectively scripted themselves parts in what followed. They spoke of Mary Paterson in erotic terms. When David Paterson entered the dissecting room and saw her naked body lying on the floor, he immediately noticed its beautiful symmetry and freshness. As it was one of his jobs to wash subjects for dissection, he would soon have an opportunity to do more than look. Paterson also said he overheard William Fergusson say that he knew this young woman. Henry Lonsdale, another of the students who was present, remembered Mary Paterson's body as one that could not fail to attract because of its voluptuous form and beauty. When she had been laid out on the

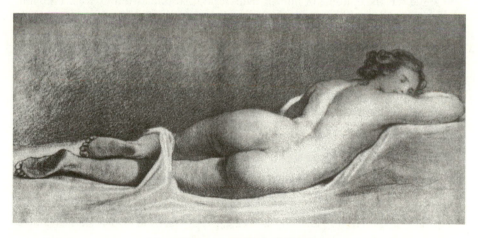

Mary Paterson, 1828, by J. Oliphant
(*Roughhead,* Burke and Hare).

dissecting table, the students crowded around her, and artists came to study her 'stretched in death, and ready for the scalpel of the anatomist'.[50]

The body was still soft when Knox received it. Some said it was still warm. Only four hours had elapsed since her murder. That is why the students and Paterson were able to lift her onto the dissecting table, which they had turned into an artist's prop, and arrange her in the way seen in the sketch. Perhaps it was while doing so that they found her hand was tightly clasping a few coins.

The sketch made by the artist, Mr Oliphant, is included in only one history of the Burke and Hare murders. It appears there as an unexplored pointer to something that can be taken for granted, but was not often articulated or sketched, about the culture of a dissecting room. But if you read about Robert Knox more deeply—what he wrote over the years on anatomy, on art, and on race—it also becomes possible to see more than a form of necrophilia in this image. The Knox revealed in these writings is a man who believed race determined everything else in life, which would have meant he saw Mary Paterson as a rare treasure, for she embodied the human form he believed to be most perfect in the world.

Knox always insisted his theoretical opinions were acquired in his daily, practical experiences. If that was so, then Mary Paterson was fruitful for his thinking in several ways, for she was a woman at the peak of the kind of beauty he was later to argue was the epitome of human civilisation, that of classical Greece. Through such women as her, Knox even linked the ancient Greeks and his own Saxon race, believing both shared a respect for beautiful women who sold their favours for money.

Mary Paterson has been sketched in a pose that is carelessly inviting. The dissecting room and all its accoutrements and inhabitants have been erased from view. She lies alone, naked before us. We are given a posterior view of her body. This was, Knox would later argue, the finest perspective on the human frame, for it presented none of the problems he believed that 'Nature had ... to overcome [to] make the front beautiful'. A posterior view hid 'anatomical organs of the most appalling shapes'. Mary Paterson has been carefully staged to display all the features Knox most admired: the 'extended and beautifully waving vertebral column' (which was, he said, nothing like 'the frightful chain of osseous nodosities supporting it'); muscles of great length and 'uninterrupted by markings or grooves' from what lay beneath the skin; and a covering 'of a velvety softness ... [that] invests and conceals', he said, 'every anatomical form'. Arranging her in this way also allowed him to hide those organs he thought were 'forms which Nature never intended should be seen'. Like his perfect form, the statue of the Venus of Melos, there was not a spot on Mary Paterson's body to indicate 'the presence of any internal organ or cavity'.[51]

It is only after looking at this image for some time that we realise she has been twisted to assume a pose difficult for a living model to hold for any length of time. Both her back and her face are presented to us. Were its subject alive, this portrait might fit into the genre of nude Venuses. William Roughhead, in whose history of Burke and Hare it is contained, says the portrait of Mary Paterson was made after the manner of the Rokeby Venus, showing both front and back by using a mirror.

There are other differences between these images. Venuses were usually posed among luxurious settings of rich velvets and sensuous silks, and in the Rokeby case, there is a cupid at her feet. In contrast, Mary Paterson is unadorned, presented as a 'natural fact'.[52] In the centre of his image, Oliphant placed what Knox called the great elevation of the gluteus maximus. From there our eyes may move in one of two directions. We might trace the curve of her vertebral column upwards—then pause in the realisation that her eyes are open (though there is no danger she will look back at us). Or we may follow the line of her legs, downwards, past the things that should not be shown, and pause at the soles of her feet, which make us hesitate for a different reason. They are dark with the dirt from the floors and streets across which she walked in life. Those feet remind us of Mary Paterson's humanity.

A narrow cloth lies across her legs at the knees, as if it has casually fallen there, or been nudged down to expose her to view. On a dissecting table, it may be that the cloth served another purpose, for while you can twist the pliable bodies of the dead into any position you like before the rigour of death makes this temporarily impossible, additional props are required to keep such a body in the position you desire. Some anatomy art reveals the extent to which a man might need to go, using pieces of cloth, ropes and pulleys, and blocks of wood to hold something that is dead in a preferred position.

I do not think that Robert Knox was indulging in a kind of necrophilia in his use of Mary Paterson, although neither am I divorcing what happened in that

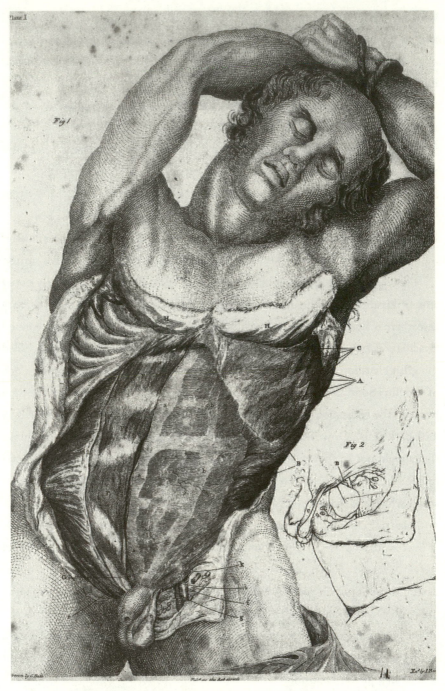

Second Stage of Dissection: Torso, 1798, by Charles Bell (Bell, A System of Dissections). *Courtesy of the Wellcome Library, London.*

dissecting room from the erotic. But Knox was a more complicated man than can be captured in just that way. He was an excellent anatomist, and yet he found the inside of the human body 'always horrible'. It was an unseemly place that lacked, he said, any 'form that sense comprehends, or desires', and he wrote of feeling in the dissecting room, with dread, his own dissolution.[53]

Reading this image of Mary Paterson in conjunction with Knox's other work, on art and race, illustrates something of the complexities at work in both the image and the man who spent his life dissecting, while refusing to turn away from the horrors it involved.

Learning anatomy was always as much about destroying as creating, which is something that many histories of medicine manage to hide. In them, it is as though the work of taking human bodies apart was a relatively simple and morally straightforward matter, but this was not the case at all. The men who went to work in dissecting rooms did so for different reasons and with different desires. These were popular places in which science combined with curiosity, punishment, art, pleasure and—always—relationships of power.

William Burke was hanged for the murders he and Hare had committed, a fate Hare escaped by providing the Crown with sufficient evidence to convict his co-conspirator. Burke was one of the last murderers to be sentenced to both death and dissection. Tens of thousands of people jostled to watch some part of those performances, and pieces of his body were souvenired during the dissection. A purse made from Burke's skin is still displayed at the Police Museum in Edinburgh.

The British parliament passed an *Act for Regulating Schools of Anatomy, 1832*, or Anatomy Act soon after the London episodes of burking. One clause of this Act rescinded the Murder Act under which those executed for this crime received the additional punishment of dissection. Instead, the new Act provided medical men with increased legal supplies of subjects to dissect. It established a system for licensing who could perform 'anatomical examinations', and where these could be undertaken. From 1832, all those in legal possession of human bodies—including the men in charge of charitable institutions such as hospitals and workhouses—could make them available for examination. The only caveats were that no person should be 'examined' if they had left instructions that they did not wish to be, and relatives could also object. In effect, all other bodies not claimed for burial within forty-eight hours of death became the surgeons' things.

As Ruth Richardson has shown, this Act was notably silent on certain matters. It did not name Britain's charitable institutions as the major envisaged source of bodies. Lessons had been learned in 1829, when an earlier and more explicit Anatomy Bill had been rejected by the House of Lords, acting in defence of the poor.[54] Nor did the 1832 Act provide a means by which people entering those institutions would be informed of their likely post-mortem fate, and so have the opportunity to leave instructions of their preference not to be dissected. No onus was placed on charitable institutions to find and inform relatives of a person's death and so enable them to claim the body. And in any event, this matter of being 'unclaimed' when you died in a charitable institution was a tricky one, at a time when to claim a body meant that you became responsible for paying the costs of its burial, which many people could not afford.

Even the word 'dissection' was missing from the Act, which used instead the phrase 'anatomical examination'. It had been deliberately chosen to disguise the actual practice that was being regulated. 'Examination' was the word generally used to describe an investigation to determine the cause of a person's death when that was uncertain. 'Dissection' was something else altogether. One London surgeon, G. J. Guthrie, made the distinction between an examination and a dissection in intimate terms when he said, 'I personally have no objection to be opened and examined as to the cause of death, but I do not intend to be dissected, if I can avoid it …'.[55] The reluctance to become a subject for dissection extended far beyond the poor.

Breaking the link between surgeon and executioner by rescinding the Act under which murderers' bodies had been subjected to punitive dismemberment meant that from now on, anatomy was to be carried out in the cause of science rather than as a punishment for criminal acts. The public, theatrical performance of such things passed from view—until Gunther von Hagens revived it in 2002.

Perhaps all we can say about such public dissections in the end is that, in both the present and the past, they are never *just* in the interests of promoting medical science. And neither, as we have seen, were those conducted in the relatively more private spaces of hospital and extramural anatomy schools.

With this knowledge of how such things were accomplished, we will now travel with medical men trained in this culture of dissecting to Britain's far-off colony, Van Diemen's Land, a place that provided them with a new set of personal and professional opportunities and constraints through which they could create authoritative identities for themselves over other people's bodies.

CHAPTER TWO
DISSECTING MARY MCLAUCHLAN

We talk about death as though it was the end of everything, but that's only the way we talk ... one doesn't slip out of life as easy as all that.

Jules Romaines, *The Death of a Nobody*, 1944

On Monday last, the execution of the unfortunate Mary McLachland alias Sutherland took place. This is the first instance of a woman being brought to the scaffold in this Colony ... the reports that have been circulated since her death appear to us almost in credible [*sic*]. We pitied the woman that could be guilty of such a crime—we now pity much more the originator of her sufferings; she has left the world penitent—may her seducer on his death-bed do the same.

Colonial Times, 23 April 1830

On Monday, 19 April 1830, to the accompaniment of the muffled bell tolling from St David's Church, a young Scotswoman mounted the high scaffold erected just inside the walls of the gaol in colonial Hobart Town, at the far end of Great Britain's world. Mary McLauchlan was dressed in a white gown, which was tied at the waist with a black ribbon. It had been especially sewn for the occasion by the gaolor's wife, Mrs Bisdee. The Colonial Chaplain, William Bedford, walked beside the young woman, Bible in hand. He was well practised in these ceremonial executions. Saving the souls of those about to hang was one part of his duties that attracted praise even from those who were normally his severest critics.

Although he was the kind of man about whom harsh anecdotes abound, Bedford was good at persuading those who were about to die on the scaffold to make the kind of public farewell the authorities most desired onlookers to witness: a full confession of guilt and due acknowledgement that the sentence of death was just, as well as some exhortation to the crowd to mend their wicked ways. Unexpectedly, Mary McLauchlan defied him. Though he had spent two days in working to save her soul, all he managed was to persuade her of two things. The first was that she should not name from the scaffold the man who had made her pregnant. The second was that she should give up the name of the woman who had assisted her to murder her infant child. Who knows with what threats of hell and promises of redemption Mary McLauchlan was made to name perhaps her only friend in this place, whom she had refused to name at her trial?

According to the editor of the *Colonial Times* (23 April 1830), as Mary McLauchlan ascended the ladder to the scaffold, she 'evinced that firmness, which sometimes is so much required by men in similar cases'. Standing above the crowd, she refused a last opportunity to confess and begged nobody for forgiveness for taking the child's life. Instead, the only voice to be heard was that of Bedford as he read from the Bible and prayed on her behalf. The executioner, John Dogherty, had never hanged a woman before. When he placed the rope around this young woman's neck, he surely took extra care with the knot to ensure a quick death, rather than an unseemly struggle while she slowly strangled. Then he drew down the hood to cover her face.

That silence from the scaffold was as powerful as any words Mary McLauchlan might have spoken under such circumstances. All that escaped her when the hangman kicked the wooden peg with his foot and the drop opened beneath her were the words 'Oh, my God' as she fell.

At what point does a human being become nothing more than an object in the eyes of those who look at her? Death itself is an untrustworthy guide to mark the dividing line. Even when we look at non-human objects, James Elkins argues, they capture our gaze, wanting something of us because they are *there* for us.[1] How much more might this be the case when what we see is a warm, soft, and disconcertingly naked body? She is a human being, like and yet unlike each of us who is gathered around her in this hospital dissecting room to watch that secondary executioner of the law, Colonial Surgeon James Scott, perform.

It is nine o'clock, and Mary McLauchlan lies in a small room filled with autumn light on the top floor of Hobart Town's Colonial Hospital. The room is crowded with men. Two or three have undressed her, removing the white gown and giving it to the executioner. It is a hangman's right to make what he can from the sale of such objects.

It is no easy thing to remove a gown from a dead body. James Scott stands impatiently by the table and its burden. He is a man with a sense of occasion and his own importance in the colonial scheme of things.

Mary McLauchlan lies, then, on that wooden table in the centre of the room, on her back. Vulnerable. She is familiar to some of the men who crowd around her, for they are acquainted with settler-colonist Charles Ross Nairne, to whose service she had been assigned as a convict woman. James Scott, for instance is, in all likelihood, about to dissect a woman he knows, for both he and Nairne are Scotsmen who have been active in this small community in establishing a Presbyterian church, which they both attend to receive lessons in how Christian men should live from Reverend McArthur.

Scott's students remove the hood that covers Mary McLauchlan's face, knowing what they might find. If death resulted from the quick displacement of vertebrae, rather than a slow strangulation, there will be no bloated purple face nor protruding and bloodshot eyes. A quick death will have made it less easy for the men in this room to see Mary McLauchlan as something other than human. If she still has the comely face that was noticed and commented upon by newspaper editor James Ross, it will make it that much more difficult for this audience to dismiss her humanity. As we have seen, a corpse has the ability to unsettle scientific proceedings. It can unnerve those who gather around it, though there are ways of making this less noticeable, and James Scott understands them.

Mary McLauchlan's eyes are open. Now that the pupils are so large, they look more black than hazel. Lying there, she seems to look up towards the faces hovering above her. Eyes open, she has not yet become a thing for the surgeons. Scott places his fingers on her eyelids and closes them. Though he is always pleased when men accept invitations to watch him perform on the dead, he has no wish to feel— even for an irrational moment—that those he is about to dissect look back. So, eyes shut.

There remains the problem of the hands. Used in life for a thousand mundane daily activities, in a dissecting room hands are poignant reminders of a person's humanity. This woman's hands have been marked by her life. One has a scar on the index finger, evidence of an accident years ago, before she was sent to this wretched place at the end of the earth. In Glasgow, those hands dressed and washed her two young daughters. On freezing winter mornings, they passed a thick plaid shawl to the cold children whose work she oversaw in Mr Dunlop's cotton factory. Her hands once passed a precious piece of paper (attesting to her good name) to a notary. In different circumstances, they murdered a baby.

It is best to avert our eyes from her feet, too. Remember Mary Paterson. Face, hands and feet, all once in motion, connecting Mary McLauchlan to those around her. Now utterly still. Waiting, while a surgeon decides which part of her he should first expose to those he wishes to impress.

Unlike a public execution, a dissection performed on behalf of the law was a relatively more private affair. Despite its didactic intent, which was to inflict terror in the hearts of those who were about to witness it, no surgeon ever set up his tools of trade and went to work on the platform beside the hangman. Instead, this punishment was always undertaken in surroundings that emphasised dissection's connection with medical science: at the Royal College of Surgeons in Britain, and in a hospital dissecting room in Van Diemen's Land.

Nor, in Van Diemen's Land, were these dissections made vicariously public. In all the press accounts I have read of executions in this colony, which is a great number, there is very little discussion of this secondary form of punishment—just a bald announcement that it had been part of the sentence pronounced. So it is difficult to understand how dissection was meant to do its work and produce terror. The few public instances in which more was made of dissection in Van Diemen's Land are notable exceptions to this rule. In 1841 Patrick Minahan shouted defiantly from the dock to Chief Justice John Pedder that while the surgeons might dissect his body, they would never dissect his soul.[2]

Another exception occurred in 1845, when Eliza Benwell became the second woman to be executed in the colony, and the Editor of the *Hobart Town Advertiser* (3 October 1845) shared with his readers a shudder of revulsion at what he subsequently witnessed in the dissecting room. It was, he said, a 'charnel house', in which portions of a recently executed body still lay on the dissecting table, 'an awful sight, over which we draw a veil'.

The third mention of a dissecting room came three years later, when the hospitals to which two bodies were carted for dissection were named, which was unusual. Perhaps the reason was that one was a private place in which the dissection would not be performed by the Colonial Surgeon.

These punitive dissections were intended by the state to be object lessons for the viewing public, but sanctioning vengeance was also a problematic matter. The emotions raised at such events could be difficult to control. In Edinburgh, a few months before Mary McLauchlan was hanged, Sir Walter Scott wrote distastefully in his diary about the multitudes who attended William Burke's execution and dissection. 'Who is he', asked Scott, 'that says we are not ill to please in our objects of curiosity? The strange means by which the wretch made money are scarce more disgusting than the eager curiosity with which the public have licked up all the carrion details of this business.'[3]

These were the words of a conservative man, but the sheer numbers who jostled to view the dissection of Burke's brain, after Professor Monro had torn off the skull cap, call into question arguments historians have made that the public viewed

dissection with outrage. The emotions that brought people to witness these performances were far more complicated than that.

James Scott had been a navy surgeon before arriving in Van Diemen's Land as the
Surgeon-Superintendent on the *Castle Forbes*, which was transporting convicts to
the Antipodes. He became the penal colony's Colonial Surgeon in 1821 and was
less controversial than the men who had previously held that position. According
to Governor Lachlan Macquarie, surgeon Matthew Bowden had been 'a man
of dissolute habits, prematurely old'; Edward Luttrell was 'criminally inattentive
to patients, [and] extremely Irritable and Violent in his Temper'; and Henry St
John Younge, an Assistant Surgeon, was 'exceedingly Ignorant as a Medical Man
… [and] almost destitute of Common Understanding, and very low and Vulgar in
his manners'.[4]

Each had gained his position more for whom he knew than for the training and
skills he possessed, which was not really surprising, as patronage was the English
way of things. In any event, no medical man with a good reputation and fine prospects of a career at home would have applied to be a government employee in a
penal colony 12,000 miles from the centre of things.

Becoming Colonial Surgeon was, however, a step up for James Scott, who was
a man of humble beginnings. And he made the most of the opportunities his new
position made available to him. He married the daughter of a former governor,
'Mad Dog' Davey, and became a wealthy man. As well as his official appointment,
Scott established a very rewarding private practice in treating the settlers who
increasingly poured into the colony. His patients included all the well-to-do people
in Hobart Town, including Governor George Arthur and his family, members of
the Supreme Court, and naval and military officers.

Those medical men who arrived in Van Diemen's Land without official appointments to the Convict Medical Department sought to make their living from private practice, but they were placed at a severe disadvantage in relation to men like
Scott, who held the kind of monopoly that made them the most *practised* surgeons

available. Men attached to the department were the only ones who had access to continuing clinical practice in the colony's medical institutions, foremost among which was the Colonial Hospital in Hobart Town. This publicly funded system of medicine was quite different from that which operated in England, where hospital medicine was a matter of charitable rather than government institutions.

James Scott and his assistant colonial surgeons also monopolised the bodies delivered to the hospital from the gallows, and these became a kind of battlefield on which Scott defended his authority when it was challenged by private medical men seeking access to the dead. The first to confront the Colonial Surgeon on this issue was surgeon William Crowther, who arrived in the colony in 1825. In 1832, when Crowther learned that Thomas Fleet and William Evans were about to be hanged for murder, he wrote to Governor Arthur to request that his students be allowed to be present at all surgical operations and dissections of subjects given over by the law for that purpose. Crowther was used to the English way of managing such things.

Arthur thought the request reasonable and forwarded it to Scott, with an annotation stating that medical practice in the colony would best be served by the liberal dissemination of the knowledge that would flow from such anatomical observations. Scott saw things rather differently. He strongly objected to what he saw as an impertinent request from a private medical man, and said it was no part of his duties to educate the pupils of such practitioners. Besides, he later expanded in a letter to Arthur, in the past he had found such 'indulgences' had been 'once, and again, abused very grossly'. Scott claimed his main concern was with 'Empyrics, and others, who improperly assume the Medical character'.[5]

But this description could not possibly have referred to William Crowther, who had arrived in the colony with excellent references from some of England's best-known medical men, including Sir Astley Cooper. And in mounting his argument, Scott was also conveniently ignoring the abuse that had been inflicted on the body of one executed murderer, the cannibal Alexander Pearce, whose skull Scott had allowed an assistant surgeon to remove from the dissecting room.

Arthur relayed Scott's angry response to Crowther, who was not a man easily subdued. He wrote again to the Governor, pointing out that if such abuse had indeed occurred in the dissecting room, then surely the proper course of action was for the malefactors to be identified and regulations enacted to ensure such behaviour could not happen again, rather than ban other medical students from watching these dissections. Scott, however, won this round of the battle.

Two years later Crowther tried again. Three men were to be executed and dissected, and he wished to ensure that his pupils would benefit from witnessing Scott's work on their bodies. This time Arthur *directed* that Scott be advised to accede to the request. However, the Colonial Surgeon was a law unto himself. He only responded to Arthur *after* he had performed his public and private demonstrations on these bodies, reporting that 'as usual' he had arranged for 'the admission of such persons as might be disposed to witness [the dissections]', which 'Several Medical Gentlemen and private individuals attended', but that

> After the Dissection, to several pupils and Medical Gentlemen of my acquaintance demonstrations were given, and various operations of surgery shown for the instruction of the young and the satisfaction of the old. I beg leave to state that I have invariably considered it my duty to make a Public Exhibition of Dissection of Criminals … ordered … to me for that purpose as soon as convenient on the day of execution, and I will continue to do so while I have the honor to be Chief Surgeon of this place … Mr Crowther and his Pupils had and have free access to the Public Dissection Exhibition, but not to be present at any demonstrations on the bodies or operations … [or to] touch the bodies without my sanctions; and as to granting indiscriminate admission for persons calling themselves Physicians, Surgeons, or Medical Students, to witness my practices in the Hospital, I will never give my consent … [the idea is] as preposterous, as were I to demand admission for my son to the offices of the Attorney General, the Colonial Architect or Surveyor General to gain from him a knowledge of their Professions.[6]

In Van Diemen's Land, as in London, bodies obtained by surgeons fresh from the gallows were particularly valuable to medical men, as they were those of people who had died suddenly, often while they were young and healthy, rather than from a terminal disease. Such bodies could be used to reveal and demonstrate human anatomy. The work Scott performed on them was, he variously stated, a matter of 'surgical operations' (1832), 'dissection', 'demonstrations' and 'various operations of surgery' (1834). They were many-layered affairs, carried out for the satisfaction of young and old, beginning with a 'public exhibition' on the day of the execution (1834).

There is no list of those whom Scott invited to attend either the public dissection or anatomical demonstrations on Mary McLauchlan's body. However, some educated guesses can be made, based on the way such things operated in England and Scott's links to other well-placed and interested colonial men. The public dissection would have been a crowded affair attended by well-connected gentlemen, as well as medical men and their students, for men attended dissections for many reasons, not only because they were learning and practising anatomy.

Reverend Bedford's son, Edward, was certainly present. He had been one of Scott's apprentices and, by 1830, was the sub-assistant surgeon and dispenser at the hospital. After he had practised for years on the dead, a report in the *Tasmanian and Austral-Asiatic Review* would soon detail the circumstances in which he performed his first operation 'upon a living subject' (7 May 1830). In addition, at least two of the hospital's convict assistants, Francis Hartwell (who had qualified as a surgeon before being transported) and John Dawson (a surgical instrument maker) would probably have been in the dissecting room. Representatives of the colonial press were invited, and amateur and professional artists would also have attended. Colonial Auditor, artist and diarist George Boyes, for instance, was interested in perfecting his representations of the human body, and had attended dissections and anatomy lessons at Charles Bell's private school in London before travelling to Van Diemen's Land. And convict artist Thomas Bock was there, for Scott made use of his artistic skills in the dissecting room, where Bock made several post-mortem portraits of executed murderers over the years.[7]

William Parramore, a young solicitor whose family's land adjoined the Nairne property on which Mary McLauchlan worked, may also have been present. He

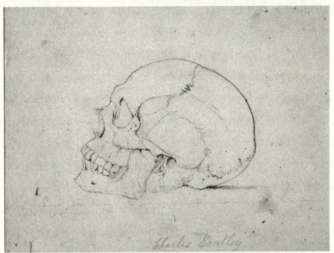

Charles Routley, 1830, by Thomas Bock (Dixson Library,
PX5 ff.20–21). Bock's first sketch of Routley was made soon
after this murderer was laid out on Scott's dissecting table. We
notice Routley's distinctive features—the thinning hair,
fashionable sideburns, thick eyelashes and well-shaped nose. The
second sketch was made after the dissection, when the skull had
been thoroughly cleaned and posed in this way for the artist. All
signs of Routley's individuality have now disappeared. Courtesy
of the Dixson Library, State Library of New South Wales.

was Reverend Bedford's protégé and had taken an interest in Scott's dissection of Alexander Pearce, writing home to tell his fiancée that the skull had been preserved by 'the craniologists', a fact that was not common knowledge at the time.

We know the order of a dissection during the early 1800s once the cutting began: first open the cavities—abdomen, thorax, and head—and only then examine the longer-lasting limbs. We have also learned about the specific kinds of uses to which women's bodies were put on the dissecting table. The only difference in Van Diemen's Land lay in the fact that a Colonial Surgeon supervised every part of this process, keeping tight control over who could witness his work or touch the corpse, while he demonstrated, operated and dissected.

Beneath James Scott's hands, Mary McLauchlan came to represent all women. First, the surface anatomy, which was a matter of observation both general and minute. Pull out the nipples to expose the excretory ducts. Separate the labia to make the external genitalia visible to all. Teach those who watch how they might orient themselves in this place, especially the medical men and their students. Insert that new instrument invented by the French shoemaker into the vagina and see how well it facilitates observation. Note the twelve signs on Mary McLauchlan's body that indicate she has given birth, including the white and shining lines across her abdomen, and the brownness of the disk surrounding her prominent nipples.

Then begin to cut. Feel the difference in the texture of a woman's skin beneath which the musculature is less developed and lies below a thicker layer of sub-cutaneous fat. Such things make women more difficult than men to dissect. Mary McLauchlan's flesh might even resist the knife, as Fanny Burney's did in 1810, 'in a manner so forcible as to oppose & tire the hand of the operator, who was forced to change from the right to the left'.[8] Let the students experience this for them-selves. Abdomen first. See what lies within. Distend the bladder, stuff the rectum to make the ovaries more visible to the eye. Search inside the ovaries in an effort to find out whether the rudiments of any future foetus lie within them, as some medical opinion asserts.

Invite the men who observe your probing to make their own discoveries in this body. Look for signs of ovarian disease, which you know is common in women, though there is presently no hope of operating so deeply on the living. Notice the way a woman's urethra curves. Perhaps this explains the dangers of using the straight catheter to which you are accustomed. Invite those who are watching you to explore the uterus, which holds the secret mysteries of the beginning of life.

Make it easier to investigate the perineum by sawing off the legs, or placing them over her head. Tell those who are admiring your work that this is the position in which they will be placed, should they need to visit you for the removal of bladder stones, as they well might. Hear their uncomfortable but appreciative laughter.

Next cut into the thoracic cavity. Pry the ribs apart and inspect the organs they normally protect. Is the heart still moving? Cut it out to observe and time the pulsations. Do the lungs exhibit the effects of the dank environment in which this woman so recently lived?

Now go to work on that all-important head, for phrenology is in its European heyday and one of your passions. You have made plaster casts of murderers' skulls. When you tear off Mary McLauchlan's skull cap, look closely for her 'organ of destructiveness', which will allow you to expound on that article in the *Hobart Town Courier* (27 February 1830) that suggested this organ may not be as enlarged in murderers as phrenologists had previously thought.

Then point out the localities of all those other human emotions and proclivities. Finally, when your listeners have been sufficiently impressed, pass Mary McLauchlan's body over to your students to do with as they please.

I am caught in the historian's impossible dilemma: longing to write this history so well that Mary McLauchlan will live again in its pages, yet knowing the belief such a thing can be accomplished is an impossible one. No historian can really make people live again. We are not resurrectionists. Nevertheless, I cannot leave Mary McLauchlan here, disintegrating beneath these men's hands and words. Like Greg Dening and Inga Clendinnen, I believe that historians can be sufficiently thorough to reconstruct something of how people in the past experienced their lives, in a multitude of contexts, which reveal something of the 'muddy actuality' of things.[9]

A DISSECTION IN REVERSE

The process of reconstruction is like a dissection in reverse.
John Gurche, Palaeoartist, cited in Casper, *The Reconstruction*

Mary McLauchlan became entangled in the early nineteenth-century British system of crime and punishment at the age of twenty-four. In 1828, she was convicted at the West Circuit of the High Court of Justiciary in Glasgow of 'theft by housebreaking'.[10] This was a charge she would dispute when asked for details of her crime upon her arrival in Van Diemen's Land.

At the time of her trial in Glasgow, Mary McLauchlan was a woman of many parts. During the daylight hours—from dawn, she said, until 'the gloaming'—it was her job to oversee the work of ten children employed to pick at cotton cloth in William Dunlop's Mill in Barrowfield Road. After dark, she was a wife to William Sutherland, a weaver, and the mother of two young daughters, aged five and three. Mary McLauchlan had been born in Saltcoats, a coastal village thirty miles south-west of Glasgow. In leaving there some time prior to 1826, she also left her family and all that was familiar, to become one young woman among many moving to Glasgow to look for work in the booming industrial city.

Whether she married William Sutherland before she left Saltcoats, or met and married him later, is unknown. He and his brothers were members of Glasgow's community of weavers, working in an industry that had been at the forefront of Britain's industrialisation and whose workers had suffered the worst kind of displacements, working conditions and poverty. By 1828, when Mary McLauchlan was accused of housebreaking and theft or, alternatively, receiving stolen goods— from the legal paperwork, it seems the prosecutor could not make up his mind— she and her family may have been in the kind of dire straits that turned so many working people into criminals, even though both she and her husband had paid employment.

Mary McLauchlan was sentenced to be transported for fourteen years. This was a harsh sentence for a woman who had not previously come to the attention of the authorities, and about whom the Tolbooth gaoler had written, 'Connexions respectable and former course of life good', though she was being 'troublesome' in his gaol.[11] In practice, that sentence of fourteen years meant banishment for life, for very few of the people convicted and exiled to the colonies were able to return home. The unusual length of this sentence might be due to the fact that she was said to have broken into a dwelling house: crimes against property were treated harshly throughout Britain at this time.

Alternatively, if we believe what Mary McLauchlan herself said about the reason for her conviction—that she had *received* goods rather than stolen them —the sentence might reveal the character of the sentencing judge rather than the system within which he operated. During that session of the Assizes, many people received harsh penalties that were generally uncharacteristic of the Scottish legal system at the time. Of the fourteen women convicted in Glasgow, half were sentenced to either life or fourteen years transportation. In summing up this session, the judge made a particular point of vilifying those who had been con-victed of receiving stolen goods, for he thought this indicated they were career criminals.

There is, however, a third alternative explanation for that harsh sentence. It could indicate that Mary McLauchlan had refused to co-operate by naming the person or persons from whom she had received those goods. In the records relating to her trial, it seems clear that William Sutherland, not his wife, was the chief suspect. Later events in the colony reveal that such personal loyalty may have been a pattern in her life.

The sentence was handed down on 22 April 1828, after which Mary McLauchlan was returned to the Tolbooth, close to her Gallowgate home, in the company of the other women who had been sentenced to periods of transportation. Some would soon become her shipmates. Five months passed before they embarked on the *Harmony*, which sailed from the Downs on the south-east coast of England on 15 September 1828. Some of that time was spent in gaol in Glasgow and the rest on the road, as the women travelled south, across the border into England.

We know very little about the logistics of that journey—of the exact route taken, the stops made along the way, whether the prisoners were given shelter at night—and even less about the kind of community that was forming between women from different backgrounds who were sharing a similar fate, though it is clear from later events that friends and enemies were being made. The women were probably manacled in leg-irons and handcuffs, either alone or joined to another, and they were accompanied by male guards. The journey was made in different kinds of conveyances, but for its final leg, through the populous southern towns of England, closed hackney coaches were used. This was an attempt to avoid subjecting the women to coarse commentary from bystanders and to ensure they had little opportunity to express the kind of bravado that left the impression they were off to a life of comparative ease in a new land.

Into Exile

Once on board the *Harmony*, the women came within the purview of the vessel's Surgeon Superintendent, William Clifford. He examined each of them personally, and they were also visited by members of Elizabeth Fry's British Society for the Reformation of Female Prisoners, which extended its work with women in prisons to those boarding the vessels that would take them into exile.

Historians have given us a good sense of life on board a convict vessel at this time. The women were divided into messes, each comprising between six and eight people, with an elected Monitor. They were allocated to berths measuring six square feet, each of which held from four to six women. These were close quarters. Then the ladies of the society distributed sewing kits which, together with the ship's allocation of Bibles and Hymnals, were meant to keep the passengers quietly occupied on the long voyage south. Some women were allocated specific tasks. One was made a schoolmistress to the children on board. Two women from each mess were required to do the laundry; some were chosen to cook. When the vessel finally sailed, it carried one hundred convict women, as well as twenty-two free passengers, seventy children and twenty-six crew members.

William Clifford made notes during the voyage. The vessel stopped at Tenerife and Rio de Janeiro, before travelling around the Cape and entering the rough southern seas to Van Diemen's Land. It was the Surgeon Superintendent's job to care for the health of all on board, as well as supervise the convict women. He was good at it, at a time when convict vessels were at last provided with adequate rations—although a ship's 'supplies' had a habit of transforming into 'cargo' that a captain could attempt to sell upon his vessel's arrival at its destination. This happened with the *Harmony*. When it docked in Hobart Town, the captain insisted that he had 'cargo' to land. Both Dr Clifford and local officials argued that it was property which rightly belonged to the colony.

Unlike some other ships' surgeons, who became notorious for their concern with the sexual behaviour of the women in their care, Clifford concentrated on the health of his charges. Mary McLauchlan is not recorded as needing his individual medical attention, but she probably suffered from the problems that commonly beset those on long voyages, such as sea sickness, constipation and dysentery, as well as the kind of difficulties specific to women, such as dealing with menstruation on board a vessel that had no provision for such events. The clothing allocated to the women is precisely listed, and no 'rags' or underclothing appear. Perhaps the women had included such items in the bags containing their few personal possessions, which they would have needed to guard against each other in a ship of thieves. Menstruation is the kind of thing that historians writing about convict women have generally ignored, although factors associated with women's reproductive health may often have been pivotal to their experience of being 'convict women'. What opportunities were there for the play of friendship and enmity in coping with such personal bodily functions during the months on board?

Mary McLauchlan was seen by Clifford as he went about his daily routine. Like other ships' surgeons, he feared an outbreak of fever in the crowded conditions on board. He kept a record of the women's health, noting in his journal that he 'experienced much good from the daily general personal examination of the Women and Children Particularly the latter'. These inspections enabled him to 'detect and assist disease in the early stages allowing them as much time on deck compatable [*sic*] with the Services as possible'. Clifford also systematically recorded the barometric pressure, the direction of the wind, and the precise temperature each day as they sailed on, first towards the Equator and then away from it again. As the ship passed through the tropics, he found it necessary to make 'free and active Use of the lancet [and] Repeated Purgatives' to relieve symptoms of 'Inflammatory fever'. There were only two deaths on board, each of an infant, one of whom had been ill when the journey began.[12]

Arrival

The *Harmony* arrived in Hobart Town on 14 January 1829, and several days passed before the women disembarked. They may not have minded the delay for, in stepping onto the shore, they were entering the unknown. From the ship's deck, Hobart Town looked like a neat but sparsely populated English village huddled around a harbour that was perhaps not unlike Saltcoats, though behind this harbour loomed the impressive bulk of Mount Wellington, which dominated contemporary sketches of the scene.

Hobart Town Van Diemen's Land, c. 1827, by George
Frankland. At the time this painting was executed, Hobart
Town had only been twenty-four years in the making. Trees
have been cleared, land cultivated, and a forest tamed. The ships
upon which the colony's residents depend are much in evidence,
as are the large buildings that accommodate the official
activities of a penal colony. No Aboriginal people are in sight.
Courtesy of the Allport Library and Museum of Fine Arts,
State Library of Tasmania.

During those first days in port, the women were visited by several Vandemonian officials bearing paper, quills and ink. Colonial Surgeon James Scott came to examine the women's health and inspect the ship. He reported that he found no infections on board, and that prisoners and ship appeared to be clean and orderly. James Gordon, the Principal Superintendent of Convicts, arrived with the Convict Muster Master, Josiah Spode, to carry out a more thorough physical inspection of the women. Each was asked about the main circumstances of her life and conviction, and her answers were checked against the paperwork that had travelled to the colony with the ship's Surgeon Superintendent.

Some twentieth-century historians have taken the information contained in this record and turned it into a general picture of convict women. They have asserted that reliable facts about the women's origins were contained in these records, since the only alternative is to believe that the women systematically lied.[13] But facts are not a simple matter of truth versus lies. There are countless matters that influence people's choice of words in responding to questions posed by officials and how they are recorded. This is a human interaction between people holding very different social positions. Assessments are made, and questions are answered in creative ways.

When the women on the *Harmony* responded to the Muster Master, in the presence of the other women and of colonial officials, certain matters would have been racing through their minds. In prison and during the voyage, stories had been told and information about the place to which they were being exiled had been exchanged. The women knew that statements they had earlier made had been turned into official records, and tried to remember exactly what these might contain. They wondered whether it would be best to restate something already said, or change it in the light of new information. They wondered about the purpose of certain questions now posed to them, and how their answers might be received, about what the listener might like to hear and what it was in their own best interests to tell him— about where the words they used might go, and what they might do there.

Add to this the emotions they experienced at this moment, such as nervousness, bravado, fear, frustration, uncertainty, anger, and even boredom. After all, there were a hundred women to be interviewed. And what was said that was not recorded? These women would not have spoken in the few sharp sentences that have been written down. Those sentences are only the way that their stories were turned into something a bureaucracy could handle. It is how messy lives are turned into a few authoritative facts.

The records of answers to questions are best seen as difficult acts of translation, rather than bald facts. It was all the kind of ritualistic examination about which Michel Foucault has written—that seemingly small matter of asking questions and noting things down, which had the effect of making subjects visible and arranging them according to a whole domain of knowledge and its concomitant, disciplinary power. A matter of capturing individuals 'in a network of writing' that described and analysed them.[14] The words checked, and sometimes changed, by Gordon and Spode identified each woman in view of future colonial need, to place her within the convict assignment system and, should she become one of the colony's many absconders, aid in her capture.

To complete the record, these men asked each woman to state briefly, but in her own words, why she had been transported and a little of her circumstances of life. This was the only time the women had an opportunity to argue with the record that had travelled with them to the colony, and many took it. Some corrected the way they were named. Mary Clark said her proper name was Potter; Mary Clifford that hers was Butler; Ellen Roberts that hers was Tomkins. Others disagreed with the gaoler's assessment of them as prostitutes: Elizabeth Daley stated emphatically '*I was not on the Town*'. Mary McLauchlan, as we know, gave an alternative to the crime that was written on her record. Mary Morgan tried to explain things to Mr Spode in a way that wrenches the heart. She informed him that she alone had been responsible for the crime of which her sister had also been convicted. This was long after there was any hope that a belief in her words would make a difference to her sister's life, though she may not have realised that.

Each woman was then questioned about her family connections. Some made sure that each family member was listed on her record, naming parents, husband and children and stating where they lived. Sophia Mendoza informed Spode that her father was Daniel Mendoza, 'the noted bruiser'. Ann Beard disputed what had been written of her marital status, saying 'I had left him'. Eleanor Sinclair placed on the record that she was married 'to Mr Charles Sinclair of Carrhness [*sic*] in Scotland' and that he was 'the cause of all my misfortune'. Then she told the

Muster Master, 'I must decline saying more just now as I do not wish to expose my family'.[15]

In the few short sentences written while she spoke, Mary McLauchlan named her husband and the fact that he was a weaver, and gave the ages, though not the names, of her two daughters. There was apparently no mention of parents, or of where her children lived now that she was so far away. In fact, she might not have known, for soon after she had been incarcerated in the Tolbooth, her husband William Sutherland had disappeared.

When all the questions had been answered to their satisfaction, James Gordon and Josiah Spode turned their attention to the women's bodies. They examined those parts that were readily available for such an inspection in order to record a picture in words for Governor Arthur's Black Books. Spode was clearly working with a pro forma designed with male convicts in mind ('Whiskers—nil'). Mary McLauchlan appears in Description Number 86. She is 5 ft $3\frac{1}{2}$ inches tall in her bare feet, and twenty-eight years of age. Examining her face for the record, Spode began with things he could see accurately for himself: complexion dark, head oval, hair brown, visage oval, forehead perpendicular, eyebrows brown, eyes hazel. Next he moved on to assessments that required more precision and, perhaps, a tool. He found nothing especially noteworthy. Most things were 'medium': the length of her nose, though it was rather flat on the ridge; the length of her chin; the width of her mouth. He asked her to open her mouth and found she had lost one front tooth on her lower jaw.

This thin description is the only one we have. We get from it no real sense of her face at all. There is a scar on the last joint of the forefinger of her left hand; but then so many of the women bore a scar of some sort that this, too, was more a norm than a distinguishing feature. Finally, after all this business with the paperwork, Governor Arthur came on board in person, as was his custom, to lecture the women on what was expected of them in his penal colony.

Assignment

Eight days had passed by the time the *Harmony* women finally disembarked. Each wore a clean set of clothing to enter the colony. It comprised a brown serge jacket, a petticoat, a linen shift, a linen cap, a pair of worsted stockings, a pair of shoes and a neck handkerchief. In addition, each carried ashore the few personal objects she had managed to retain through prison and voyage (except for Elizabeth Taylor, that is, who had somehow managed to fill and hold onto a trunk full of possessions).

The women were rowed ashore in small boats at 4 a.m., in an attempt to avoid the kind of hullabaloo caused in a colony of men when 100 women arrived. Then they were marched one mile out of town, along Macquarie Street towards Mount Wellington, beneath which stood the high stone walls that enclosed the Cascades House of Correction, or Female Factory.

Over the following days many of the women were assigned to settlers and officers who had applied through the Superintendent of Convicts for household servants. Some went to work for men who were, within a year, to be intimately involved in discussing Mary McLauchlan and working on her body and soul: the muster master Josiah Spode, chaplain William Bedford, Colonial Surgeon James Scott, newspaper editor Dr James Ross and magistrate and lawyer Joseph Hone.

Mary McLauchlan was assigned immediately to Scottish settler Charles Ross Nairne and his wife Katherine Stirling, who lived on a property they had named Glen Nairne, in the vale of the Coal River, north-east of Hobart Town. Those requesting convict servants had some choice in the women they obtained. There were even complaints that settlers boarded the vessels while the Principal Superintendent of Convicts was taking down descriptions. Perhaps it was Mary McLauchlan's Scottishness that appealed to Nairne, a merchant who had emigrated from Paisley, just south-west of Glasgow, with his wife and their young son in 1822.

Against critics who argued that transportation was not sufficiently dreaded by those committing crimes in Britain, Governor Arthur was supportive of the assignment system. He pointed out that

> Deprived of liberty, exposed to all the caprice of the family to whose service he may happen to be assigned, and subject to the most summary laws, the condition of the convict in no respect differs from that of a slave, except that his master cannot apply corporal punishment by his own hands or those of his overseer, and has a property in him for a limited period only.[16]

As Kay Daniels, Joy Damousi and others have argued, the experience of this system was also a gendered one.[17] While convict men were mostly required for out-door work that was, to a certain extent, governed by the hours of daylight and the season and so had a finite beginning and end, convict women worked as domestic servants and, like such servants everywhere, they were placed in a more intimate and open-ended domestic relationship with those they served. One effect was that they could be vulnerable to expectations of and demands for sexual favours. Another, that they were seen as a sexual threat to the family. Some, no doubt, played the game of sexual politics themselves.

Ellen or Martha, c. 1824, by George Boyes. Boyes wrote of this sketch, 'It is a picture of the Female Servant, I mean the very best of them' (*letter dated 5 November 1824, Chapman,* The Diaries and Letters of G. T. W. B. Boyes). *There is no sign here of the insolent, sluttish behaviour of which so many convict domestic servants were accused by settler-colonists. Courtesy of the Tasmanian Museum and Art Gallery.*

Many convict women, it seems, had been domestic servants prior to being transported. They would have been familiar with middle-class household dynamics and understood something of the social relationships in which domestic service took place. In the colony, the relationship between masters and their assigned servants was, at least formally, a regulated one. While at this time no masters needed to pay their assigned servants for the work they undertook, they were required to supply them with such things as food and clothing. However, as is always the case, there were ways around the rules, and some masters were better than others. It seems to have been common to neglect to supply the requisite provisions to the convict women who became domestic servants.

Mary McLauchlan may initially have felt comfortable in her placement with the Nairnes, who shared her country of origin and spoke in a similar accent. However, she was not familiar with the role of a domestic servant. The words on her convict record, which place her as being one, are wrong. So although, as a woman with a family, she knew how to cook and clean and undertake all the other tasks that family life required of working-class women at this time, the ins and outs of living in a well-to-do family as its domestic servant were new territory for her. Besides, even those who were familiar with the performance of such tasks in exchange for money would have found assignment as domestic servants in Van Diemen's Land different. In this colony they were *convict* servants, members of a despised group. Just see how the settlers spoke of them. Elizabeth Fenton said they were an 'immoral physical force', and complained that her child's nurse was a 'vile drunk'. John West thought they were 'woman deprived of the graces of her own sex, and more than invested with the vices of man'. Robert Crooke said convict women turned settlers' houses into 'veritable brothels'.[18]

At the same time, *as* women, they were rarities in a colony that was overwhelmingly composed of men, both convict and free, and this surely had an effect on the workings of sex in such a place. When these women were unhappy in the homes to which they had been assigned—for whatever reason, from being bored to being victims of violence—there was no legal way in which they could remove themselves, unless they were sufficiently brave, confident or foolish to make a charge against their master before a local magistrate, whom the master would know and with whom he would mix socially. Such complaints *were* made, but it was extremely rare for the women who made them to be believed. While Governor Arthur was concerned with the behaviour of those who gained assigned servants, and spoke of taking them away from morally reprehensible men, in practice this seems rarely to have happened.

Within two months of going to serve the Nairnes, Mary McLauchlan was pregnant. We do not know under what circumstances, whether it was a matter of rape or of sudden mutual desire (though there was nothing mutual about the master–servant relationship), or of sex offered as a gift in the hope of something being received in return. Nor, given Reverend Bedford's perseverance over the weekend that preceded her execution, do we know for certain that Nairne was the child's father. All we know is that newspaper editors *did* know the name of the father of her child, which was the subject of gossip in Hobart Town. They expressed shock, for he was an educated man much 'above' Mary McLauchlan in station, and had refused to take responsibility for the child.

The military men on the jury that convicted her of murder thought that she was not beyond this man's reach when she was incarcerated in the Factory. They believed he could have influenced her to murder the child. The other women in the Factory overheard her saying she wished the child would be born dead. And in reporting her execution, the editor of the *Tasmanian and Austral-Asiatic Review* (23 April 1830), who had a soft spot for women, named in capital letters the man to whom she had been assigned when she became pregnant, 'C. R. NAIRNE'. This was not that newspaper's usual way when reporting that a master's servant had committed a crime.

These matters are too circumstantial to be considered evidence that Nairne fathered the child, though no more circumstantial than the evidence on which Mary McLauchlan had been convicted before the Glasgow Court of Justiciary two years earlier. With the benefit of hindsight, there are other factors that point in Nairne's direction. His relationship with his wife seems to have fallen apart soon after Mary McLauchlan was executed. Letters indicate the couple to be living apart and, though the family moved to Launceston a few years later, Nairne soon left, alone, for Sydney, where he died in 1842. Moreover, the Nairnes' assigned servants frequently absconded. Add to that the fact that Katherine Stirling was eight months pregnant when her husband arrived home with the family's attractive new servant, having travelled with her from Hobart Town to that isolated place.

Mary McLauchlan was pregnant at Glen Nairne for five months before there is any sign in the records of something being wrong there. At five months it may be that others in the household noticed the pregnancy. On 10 August 1829 Nairne took her from Coal River to Hobart Town to the office of the Principal Superintendent of Convicts, now Josiah Spode, and charged her with an unspecified form of 'misconduct'. It may have been the kind of trumped-up charge masters commonly made when wishing to rid themselves of troublesome servants. Whatever the accusation, Mary thought it was unfair. She spoke up for herself, telling Spode that the Nairnes had not given her the 'proper' amount of clothing that was her due. Perhaps, with her expanding belly, it was the issue of clothing that had precipitated this trip to Hobart Town, or a wife's fury on noticing the need. Or perhaps this was one of the few charges a woman *could* make against the family to whom she was assigned that might count for something in the scheme of things, since a failure to supply the proper allowance of clothing meant the Nairnes were breaking one of Governor Arthur's recently proclaimed regulations.

Josiah Spode remanded Mary McLauchlan to the House of Correction and told Nairne that he wished to hear what his wife Katherine had to say about the matter. Handing out clothing to servants would have been her responsibility. This directive from Spode meant that Nairne had been ineffective in persuading the Principal Superintendent of the rights and wrongs of the situation and thus ridding the family of this servant. The convict clerk who was recording the proceedings wrote on Mary McLauchlan's conduct record that they would hear from Katherine Nairne on the following Monday, which indicates she had not travelled to Hobart Town with her husband and the convict woman. Then, thinking this was the end of the interview, the clerk closed off the record in his usual way, by writing 'P.S.' for Principal Superintendent at the end of the sentence. But something else was said, and it made him take up his pen again. Perhaps Nairne hesitated, reluctant to obey Spode and go to get his wife. Perhaps Mary McLauchlan spoke quickly, before Nairne had a chance to leave the office, knowing she could expect no support from Katherine Nairne.

Whatever the subtleties of the situation, Mary made another, unnamed charge against her master and mistress. The clerk wrote only 'when investigated is found to be witho. any foundation', though the only 'investigating' Spode could have undertaken on this day was to ask Nairne whether or not Mary was telling the truth. The next words written on her record show the result of that. It is noted that she had left Nairne's house without permission on the previous Saturday, and now was to be placed in a cell on a diet of bread and water for six days at the House of Correction, following which she would join the 'C', or criminal class of women—the most serious offenders—before being reassigned far from Hobart Town, into the 'Interior' of the island.

The severity of this sentence tells us something. Mary McLauchlan had not previously been charged with any form of misconduct in the colony, and first offences were usually treated quite leniently. Even the women who were constantly absenting themselves from their places did not receive such a severe sentence. And the punishment for becoming pregnant while assigned was six months imprison-ment in the crime class. Only the most unmanageable recidivists were assigned in the interior of the island. So this was done to keep her quiet. And there was no further talk of hearing from Katherine Nairne.

Was it rape, or was it abandonment? Could Mary McLauchlan have naively hoped that Nairne and his wife would allow her to stay in their family and raise her bastard child so close to their own children? Trusting in the willingness of others to save her in the face of terrible odds was something of a pattern in her life. In defending herself against the charge of theft in Scotland, she had trusted a relative of her father's to give her an alibi that would have had the effect of making this witness the subject of investigation herself. The one person she placed no such trust in was her husband, William Sutherland.

From this time on, Mary hated both the child she carried and its father, and did not care who knew it, to the point of making statements that would see her hang.

In the House of Correction

In becoming pregnant, Mary McLauchlan had committed an offence that had been specifically codified in the colony to apply to its convict population. Pregnant women were sent to the House of Correction which, like the bridewells in Britain, was an institution that served more than one purpose. It was a gaol, a place that housed women awaiting assignment, and also one in which women lived when they were returned as being 'useless' to their masters for reasons of pregnancy or because they had an unweaned child.

Within the system of classification that Governor Arthur had established in 1829, attempts were made to separate the Factory women into three distinct classes, each distinguished by the clothing worn and the work undertaken. In practice, however, it was difficult to patrol the borders between the classes, for in these early years the Factory comprised only one yard surrounded by cells, and the convict women could not easily be separated from each other. Nor were they separated from those employed at the institution, who could always be persuaded to act as a conduit between the inside and outside worlds.

In constructing the institution's regulations, which were published in the *Hobart Town Gazette* on 3 October 1829, Arthur had paid minute attention to every aspect of the lives of the women who were to be imprisoned there. He decided that their clothes would be made of 'cheap and coarse materials', and that their daily supply of food would consist of bread, gruel and soup. He made rules to govern their behaviour and rules to determine the behaviour of those who worked in the place. The Factory's Superintendent was Esh Lovell, whose wife, Rachael, served as its Matron. Both would give evidence at Mary McLauchlan's trial.

Esh Lovell had the power to punish those women who disobeyed orders, neglected their work, used 'profane, obscene, or abusive language', or were 'turbulent, or disorderly, or disrespectful' in their conduct. He could incarcerate such women in 'a dark or other cell' until the case could be heard by the Principal Superintendent of Convicts, who could, in turn, prescribe a range of punishments, including the one the women were said to hate the most, which was the removal of their hair. In April 1830, the newly appointed Principal Superintendent, Roger Woods, had this sentence carried out. Then he burned the woman's hair before her eyes for good measure. It was a sufficiently controversial act to attract comment in the press. Having the power to do such things was probably a source of pleasure for this man, who seems to have been a misogynist. He was a drunkard and beat his wife.

In practice, however, the harsh regime was not always carried out. Some women imprisoned in the Factory had carved out a kind of power base of their own, which they enforced by tactics of intimidation and violence. This means that for other women, the Factory must have been a place in which they often tried hard not to be noticed, for they could be the victims of other inmates' abuse.

In addition, as in any institution, there were ways to subvert the rules. Most importantly, the women there talked. Conversations, arguments and gossip took place and were overheard, everywhere: within each class; from hammock to hammock after the women were locked into the cells at night; and during the day, across the space between the washtubs. Between the classes, in the corridors that linked cells, nursery, privies and hospital. Between prisoners and the free women who worked in the Factory, as goods and letters were transmitted. Women who were experienced in the systems of Factory and assignment took pleasure in informing newcomers about what they could expect. The names of certain settlers circulated, as did strategies for dealing with them.

During the months Mary McLauchlan spent in the Factory, she shared it with some of the most violent, and therefore most noticeable, women in the colony— the kind about whose disorderliness twentieth-century feminist historians have written. Others are almost invisible to us because they caused no such outrage and mayhem. Under ordinary circumstances, Mary McLauchlan would have been one of these. Some of the women in the Factory at this time were known to her. A few had been her cell-mates in the Tolbooth during all those months awaiting transportation, and on the journey south to the Downs to board the *Harmony*. One of these was Mary Cameron, a woman more practised in crime than Mary, who was to play an important role in all that followed.

Others had become known to her on the ship, including Sarah Bromley, who had also arrived back at the Factory from assignment on 10 August. Perhaps the two women had met that day at the Superintendent's office. Unlike Mary McLauchlan, Sarah Bromley had arrived there alone, having taken Governor Arthur at his word and come to report being 'ill-used' by the man to whom she had been assigned. She actually expected something to be done about Dr William Bohan of the 63rd Regiment.

Also present in the Factory while Mary McLauchlan was there were some women she probably met because they were placed on assignment in farms surrounding Glen Nairne. One, Mary O'Donnell, was sent to the Factory soon after Mary had arrived there. This woman would have been the subject of much gossip, for she had recently been tried in the Supreme Court for the wilful murder of her newly born child, and acquitted.

After six days in a solitary cell, Mary McLauchlan was set to work among the other women of the crime class. She spent her days standing in the Factory's dank courtyard bent over a stone trough, washing 'for the Establishment, for the Orphan Schools, [and the] Penitentiary', and labouring at 'Carding Wool, Spinning, or in such other manner as shall be directed by the Principal Superintendent'.[19] This included unpicking the coils of rope used on ships and washing out the tar that impregnated them, a job that made hands bleed. As they worked and talked, the women in the yard rarely saw the sun. They were surrounded by two storeys of stone walls, cells and work rooms, behind which lay high terrain. It is difficult to understand how James Scott and his fellow magistrates could have recommended this place to Governor Arthur as the site of the colony's new House of Correction two years previously. The recommendation was especially odd for a medical man who held the view that ill-health was caused by the kind of miasma that developed in stagnant places. Soon Scott's former pupil, Edward Bedford, would be blaming the site for the deaths of young children housed in its nursery, though the women whose children they were knew better than that. The children's deaths were due more to being taken from their mothers, to poor rations, and to neglect by so-called 'nurses'.

It was winter when Mary McLauchlan, five months pregnant, began this work. She stood ankle-deep in cold mud when the Rivulet that passed by the Factory overflowed on its way to supply water to Hobart Town. The water also seeped into the cells in which the women slept and into those other cells in which some were isolated day and night. It was all very well for British-trained medical men at this time to be instructing their students about the best way to manage a pregnancy—through rest and tranquillity, loose clothing and good diet. Such things, unlikely for most pregnant women in the early nineteenth century, were impossible in such a place as the House of Correction.

A Birth and a Death

Early in December 1829 Mary McLauchlan's labour began. We do not know much about what happened during the following hours, or what she thought about in between the spasms of violent pain. She may have remembered conversations about new-born child murder with Mary O'Donnell, who had previously given birth to an illegitimate child. She would certainly have known about what was likely to become of the babies born to convict women in the Factory, where the infant mortality rate was extremely high (208 out of 794 between 1830 and 1838).

Soon after her son was born, his body was found in one of the Factory's privies under circumstances in which there was an element of doubt about whether his mother had killed him, or he had died a natural death.

No official records relating to the trial exist, though newspapers reported it. They note witnesses stating they had often overheard Mary McLauchlan wish 'that

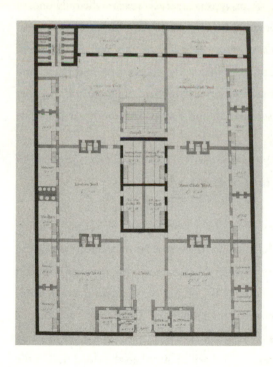

Ground Plan of the Female Factory at the Cascade, 5 September 1827, by John Lee Archer (AOT PWD/266/389). This plan shows the separate yards for the different classes of women. The Crime Class is at the back of the Factory, near the solitary cells, the workshop and the chapel. Mary McLauchlan's baby's body was found in one of the privies opening onto the Crime Class yard. Courtesy of the Archives Office of Tasmania.

the infant she bore might not be born alive'.[20] But wishing and acting are not the same thing. We are left wondering about a number of things. If it was true that Mary wished the baby dead, why did she wait until he was born rather than abort him much earlier in her pregnancy? There were ways, then as now, and women knew them. And since she pleaded that she had not killed the child, on what evidence was she convicted? How was the decision made that this death was a matter of deliberate intent rather than stillbirth or an accident, when we know the kind of doubts that plagued nineteenth-century medical men in such matters?

To prove it was murder, Attorney-General Algernon Montagu had to remove all reasonable doubt that the baby had been born dead, or died of natural or accidental causes soon afterwards. Yet such things were notoriously tricky to pin down —unless somebody other than the woman giving birth had witnessed the event and gave evidence that it was murder.

The fact that Mary McLauchlan chose to give birth in the privy, rather than the room at the Factory that had been designed for such a purpose, would have told against her plea of 'not guilty'. A privy was a *private* space—the only one, in a prison. In a privy, such a murder could be disguised. We will never know what happened in the privy in the Female Factory that day, though we can be sure of some things. That small, dark place was filled with terrible pain and panic, and Mary McLauchlan's friend, Mary Cameron, assisted her, while outside the door other women engaged in excited speculation and gossip. Somebody carried a tale to Esh Lovell or his wife.

Trial for Murder

An inquest followed the discovery of the baby's body. Like everything relating to the law's dealings with Mary McLauchlan in Hobart Town, in great contrast to the legal records surrounding her original trial in Glasgow, very few words remain of the proceedings. There is just a note, 'N.B.', on her convict record which states that Joseph Hone acted as the Coroner at the inquest, on 3 December, and that he committed her for trial on a charge of murder. We hear nothing further of Mary McLauchlan until four months later, on 15 April 1830, when she stood trial in the Supreme Court for the murder of her bastard male child.

Chief Justice John Pedder presided over the court, which had only been in existence since 1824. He had been sentencing people to hang for a wide range of crimes ever since. Juries were made up of seven military men, for in this penal colony the concept of trial by peers was a problematic one. As was common at the time, no barrister defended Mary McLauchlan, although court records indicate that in Van Diemen's Land, as in England, some murderers did have such representation, and it could make a difference.

According to the *Colonial Times* (20 April 1830), a 'vast number' of witnesses spoke at the trial. The *Times* reported that the investigation was 'most … patient … lasting from the morning until late in the evening'. This could, however, have been due to the fact that the court was a notoriously inefficient institution. Chief Justice Pedder complained that it was usual for the gentlemen of the jury to fail to arrive until well into the morning.

The most important witness to speak at the trial was probably Assistant Surgeon James Bryant, who provided the medical evidence conducive to a finding of guilt. It was his job to visit the women at the Factory every day, and we know from the record of a quarrel he was about to engage in with his superior, James Scott, that he performed his duties conscientiously. Bryant had complained that Scott failed to fill his requisitions for medicines for the women at the Female Factory. Ever on the lookout for slights to his authority as Colonial Surgeon, Scott responded by denigrating Bryant's qualifications for the position he held, though his main complaint against the junior surgeon was that 'On several Saturdays, the day appointed by me for visiting the Female Factory, he (Mr Bryant) showed great want of attention in not being there to receive me … Mr Bryant has acted with a degree of indifference towards me as head of the Department … My chief complaint is his appearing to wish to act independently of me.'[21]

Nevertheless, Bryant's comparative lack of experience and skill could have been important to the outcome of Mary McLauchlan's trial. He seems not to have informed the court that evidence in such cases of murder was a matter of great debate. In London, William Hunter, a much better qualified medical man, had earlier pointed out just how easy it was to misread an accidental death as 'murder', and Hunter regretted that many in his profession were 'not so conversant with science as the world may think; and … are a little disposed to grasp at authority in a public examination, by giving a quick and decided opinion, where it should have been guarded with doubt'.[22]

James Bryant was the only medical man to appear as a witness at Mary McLauchlan's trial, in contrast to the earlier trial of Mary O'Donnell for a similar crime. In fact, the differences in the treatment of these two Marys reveal something about the workings of justice in Van Diemen's Land at this time, and the pivotal role a woman's assigned master might play in it. Perhaps Mary McLauchlan bore aspects of Mary O'Donnell's story in mind while making her own decisions in that place.

Mary O'Donnell had also become pregnant while on assigned domestic service, and the differences in the treatment of these two women began with their masters' responses to that fact. Captain Cooling, who was not the father of Mary O'Donnell's baby, allowed his servant to stay on his farm during her pregnancy. It was not far from the Nairnes' property. Soon after Nairne took Mary McLauchlan back to Hobart Town to be rid of her, Mary O'Donnell gave birth to a male child and the baby was found dead in her apartment. O'Donnell was taken into custody by Gilbert Robertson, the Chief Constable at Richmond, but the surgeon in attendance, Robert Garrett, certified that she was not fit to leave the farm, nor could she attend the inquest over her child's body, being too ill.

The Coroner, Thomas Lascelles, who heard what witnesses had to say about this birth and death, was obviously familiar with the difficulties inherent in deciding between 'accident' and 'murder' in such cases. He was careful to inform the jury about the kind of evidence a higher court would require. Lascelles asked them to consider whether the child had been born living, whether the pregnancy had been concealed, or whether there had been wilful neglect of any opportunity to preserve the child's life. If so, this became a matter of murder. On the other hand, if the jury thought Mary O'Donnell had been 'seized with premature and unexpected labour' and had no opportunity to obtain assistance, no murder was involved.[23]

At the inquest over the baby's death the jury, having heard from surgeon Garrett, decided this was likely to have been a case of murder and the matter was sent to trial in the Supreme Court. However, during the six months that passed before the case was heard, Mary O'Donnell stayed on at her master's farm where, according to the *Courier*, she was given 'humane treatment from Mrs Field and Captain Cooling'.

At the trial, Robert Garrett again appeared as a witness. Then, on the second day, a more authoritative medical man, in the form of Colonial Surgeon James Scott, was asked for his opinion. Had Captain Cooling arranged this intervention? In addition, a third medical man, William Bohan of the 63rd Regiment—he who was then ill-treating his own assigned servant, Sarah Bromley—sat as a member of the jury. Mary O'Donnell was found not guilty of murder.

The differences of opinion between these medical men must have muddied the waters sufficiently to make the determination of cause and effect no easy matter for this jury to decide. It may have made a critical difference that James Scott had been trained in Edinburgh, where every medical man learned and was examined on medical jurisprudence, in contrast to the situation in England. In contrast, one year later, what the less qualified surgeon, James Bryant, said at Mary McLauchlan's trial established in the minds of the jury that there was no doubt she had murdered her baby.

He probably talked of what he found on the baby's body, retrieved from the privy, to indicate an unnatural death. If by strangulation, he would have seen a windpipe filled with mucous, a red and livid countenance, a swollen tongue, a red line about the neck. If by suffocation, extraneous matter would appear in the cavities of the baby's nose and mouth, there would be a 'falling in' of flesh about the pit of his stomach, as well as blood in his lungs and the cavities of his heart, distended veins about the head, froth about the mouth, and more.

In contrast, a stillbirth was indicated by a dissection that revealed a brain which 'appears fluid like water', thick coagulated blood in the heart and other blood vessels, soft and flabby skin that was red or scarlet throughout; putrifaction, an umbilical cord devoid of humous, and relatively soft bones of the skull.[24]

Perhaps no other medical men were called to speak at Mary McLauchlan's trial because the evidence of murder was overwhelming, although, as Catherine Crawford has shown, at this time it was believed that a medical man who gave an opinion in a court of law needed to be well informed in 'the whole of medical science'.[25] James Bryant was not. From the sightings we gain of this man over the years, he seems to have been a compassionate yet pugnacious and foolhardy individual. Elizabeth Fenton, wife of a well-to-do settler, spoke well of his care for her baby. She credited him with saving the baby's life in 1829, and contrasted his attitude and skills to those of the senior surgeon she had first consulted, which was probably James Scott, given the family's status. Elizabeth Fenton believed the senior surgeon had abandoned her baby to an early grave.

But Colonial Auditor Boyes captures another side of Bryant. Writing on 19 June 1830, he said '[Bryant] is a pleasant fellow but/oh the but/he seems to have resolutely made up his mind upon difficult points that are considered by many still open to controversy and is perhaps slightly tinctured with dogmatism'.

Bryant was also a man who became foolish in his cups. Boyes gives us a memorable, Pickwickian image of the surgeon, who had arrived at Boyes's house:

> in the evening, well refreshed. After tea a glass of whiskey almost annihilated him. He got up and addressed 'Ladies and Gentlemen' a dozen times, but although he tried as many different subjects he could get very little further than the first full stop.[26]

Mary McLauchlan's trial was conducted before Pedder and a jury comprising seven men of the 63rd Regiment (captains Pery Baylee, Thomas Patterson and John Mahon, as well as lieutenants Christopher Dexter, Henry Croley and Robert Cart, and Ensign John Montgomery). Baylee acted as foreman. He was soon to join surgeon Robert Garrett at Macquarie Harbour, where he became its unusually humane Commandant, a protector of Aborigines. These men were also bound in social relationships with others involved in this trial. Baylee played whist with Patterson, Croley, the Sheriff Dudley Fereday, and the Solicitor General, Alfred Stephen. Mahon had been close to lieutenant-governor Davey, and so would have known James Scott's wife, who was Davey's daughter. All but Cart and Montgomery, the lowest in status, were friends who dined with Boyes and other well-connected colonists.

Two of them had also served as jurors at Mary O'Donnell's trial, and found her innocent. But at the end of her trial, Mary McLauchlan was found guilty of murder and sentenced 'to be hanged by the neck [until] she would be dead on Saturday next the 17th inst. at the usual place [unreadable] and when dead her body to be dissected'.[27]

Aftermath

Over the weekend following the trial, after Pedder had pronounced the sentence, Governor Arthur's Executive Council met to ponder the case, as was usual for capital offences. Arthur's thoughts are not recorded in the Minutes of these meetings. Indeed, he seems to have taken a remarkably passive role. He may have been ambivalent about this particular conviction, for he was as much concerned about the morality of the settlers as with that of the convict population, and the fact that a settler had committed the act that resulted, nine months later, in a murder, would have concerned him greatly.

The meetings of the Executive Council could be uncomfortable. According to Arthur, all of the men who sat on the council were thoroughly engaged in work connected to their colonial positions and so felt it an 'intolerable burden' to attend council meetings—though this may just have been the Governor's way of pointing out to the British Colonial Office that the council was a burden all round. Four men sat on the council at this time: Arthur himself, Chief Justice Pedder, the Colonial Secretary John Burnett and the Colonial Treasurer Jocelyn Thomas. As Arthur noted to William Huskisson in the Colonial Office in London, these men had 'their various fancies, and divergent views' and the establishment of the council had blurred the line around 'where the Governor's authority extends', which was not 'desirable in a Penal Colony'. He said it had therefore been his system 'to yield in every *doubtful* matter, [rather] than attempt to try the extent of an influence which, in the event of resistance, might prove of questionable authority'. Pedder and Burnett disliked each other. Arthur, however, thought the members of the Council 'well informed and strictly honorable men, and, by far, the best qualified in the Colony' and said he was on good terms with them all.[28]

The Executive Council met for four and a half hours on Saturday and again on Sunday for half an hour to discuss Mary McLauchlan's sentence. This was unusual.

These meetings provided an opportunity for an assigned servant's master to make a difference to the judicial outcome of a case; following other trials, sentences were commuted when masters spoke on their servants' behalf. Discussions about Mary McLauchlan reveal the differences between council members. At the first meeting both Jocelyn Thomas and John Burnett recommended that her execution be delayed. They expressed concern that the crime seemed to have been carried out without 'any known adequate motive' and they wished to find out whether or not the father of the infant, 'who was supposed to be a person of better education and higher rank in society than herself', had incited Mary McLauchlan to commit the murder. They hoped that 'something might come to light extenuating the enormity of the crime … which might avert the distressing necessity of the first public execution of a female in this Colony'.

Thomas and Burnett were compassionate men (and each was soon to be charged with nefarious dealings where money and land, respectively, were concerned). Jocelyn Thomas had done his best for a young girl in the Orphan School whose assets were being seriously mismanaged by magistrate Joseph Hone, to such an extent that she would not be able to escape from that wretched place. John Burnett was said to be a true gentleman, 'of the old school'.[29] Yet Chief Justice Pedder had no doubts about Mary McLauchlan's conviction. In the Executive Council, he insisted that he could agree to no delay or any further investigation. He said he 'could not advise the Lieutenant Governor to interfere with the course of the law' (that is, with the sentence he had already pronounced).[30] In other cases, Pedder *did* condone interference with the law.

In the face of this conflicting advice from members of his Executive Council, Arthur prevaricated. He informed them that he was uncertain about whether such 'interference' took place in England in similar cases, and that he would respite 'the unfortunate woman' until Monday, on which day she would be executed if no further evidence came to light. Who knows what he expected might happen between Saturday and Monday, for he set no investigative process in train until he received a letter from Pery Baylee on the following day.

When the Executive Council met again, soon after some of its members had listened to Reverend McArthur preach on redemption from sin, Governor Arthur informed them of the content of this letter. It had been written on behalf of the jury to beg for mercy for Mary McLauchlan. Arthur had it in his power to grant that wish, for in the colony he stood in place of the King where the granting of mercy was concerned. Baylee informed Arthur that the jurors believed Mary McLauchlan had been 'driven to commit the crime by a sense of degradation and shame from the fear that the birth of her child would become known to her relations at home'. They had also been impressed by 'her deportment at the trial [which] was submissive and resigned'.

Arthur informed the council that when he received this letter, he had sent Joseph Hone to visit Mary McLauchlan in her cell 'to take any statement which the prisoner might have to make'.[31] In this way, through Hone, Mary McLauchlan was given an opportunity to save her life by placing the blame for the murder on the influence of a man she was said to hate. She refused to do it. Instead, Hone reported back to Arthur that she told him nobody had influenced her actions in any way at all. So much for the popular view that this woman had murdered her child as an act of vengeance against its father.

But can we believe Joseph Hone's version of what she said? Hone shows up everywhere in this case. He was a lawyer admitted to practice in the Supreme Court in 1824, soon after arriving in the colony as the new court's Master. He was connected to Nairne: they were two of the seven directors of the Derwent Steam Navigation Company. He had also acted as the Coroner at the inquest into the death of Mary McLauchlan's baby; he was present at her trial as Master of the Court; and now here he was again, given sole responsibility for offering her an opportunity to say something that would enable Arthur to extend her mercy—and save her life.

Everything depended on the way Hone presented this offer, and on whether she understood that that was what it was. Yet Arthur could hardly have chosen a man less likely to do it well. Hone had so many mental and physical quirks that Mary McLauchlan could well have misread the opportunity that was being offered to her. Historian Peter Chapman notes that he was considered to be 'little removed from an idiot'. He drew large crowds to the court to watch his antics, which included pulling grotesque faces before proceeding to judgement, and waggling his fingers in the air.[32]

Mary may also have been afraid of Joseph Hone. As the Coroner who sent her for trial, he had already shown his belief in her guilt. Nor was he a man who could be trusted where the vulnerable were concerned. Boyes was soon to write in his diary that he had lost all faith in Hone's truth, sincerity, honesty or humanity. 'I have no belief whatever in the goodness of his heart and hold his principles and rules of conduct at a very low price.'[33] Unfortunately, Boyes is not specific about the reason for this loss of faith, which happened over the months between Mary McLauchlan's committal to trial and her execution.

So, there was no sensitivity or compassion there. In addition, Hone toadied to those above him in social position to such an extent that injustice resulted. In 1836 he would sit as one of two magistrates in a trial in which an inhabitant of Hobart Town who had 'respectable connections' was accused of assaulting his assigned servant so badly that she died, though not before naming him as the man who had attacked her. Evidence of her death-bed charge, and that of the surgeons who attended her before death, was not produced in court; and the magistrate at Launceston who took her statement, a Mr John Clark, refused to attend the inquest. According to the *True Colonist* (5 February 1836), in his charge to the jury, Hone made a 'crying, whining appeal to the accused, and his friends' and talked of his own intimacy with the accused. The murderer was convicted, but he received no sentence.

Mary McLauchlan's life lay in this man's hands. He reported back to Governor Arthur that 'the prisoner had declared to him that no person outside the House of Correction had endeavoured in any way to influence her to commit the crime'. Jocelyn Thomas, who thought little of Hone's worth, refused to let the matter rest. He mounted another argument, urging clemency on Mary McLauchlan's behalf for reasons that related to larger debates about crime and punishment in the British world. Executing a woman in this community, Thomas argued, would not have the desired deterrent effect for two reasons. First, because her crime was one of 'rare occurrence'; and second, because no further purpose could be served if she 'suffered the extreme penalty of the law', since the delay between sentence and execution had excited 'great interest and strong feeling in the minds of the inhabitants, so much so that executing her now could not produce the same anxious, deep-felt sentiments of mingled pity and terror which at present pervade all classes'. He said the delay itself had been exemplary. Indeed, he argued that without the delay Mary McLauchlan would have 'gone to her grave and her crime would hardly have been known'.[34]

Execution

While Jocelyn Thomas continued to argue valiantly for her life, in the cell close to the place where she would hang, Mary McLauchlan was being 'assiduously attended' by representatives of both the Church of England and the Presbyterian Church, in the persons of Reverend William Bedford and Reverend Archibald McArthur. Such reverend gentlemen were vital to the ritual of public executions. Through them the authorities tried to ensure that those witnessing executions were provided with a lesson in morality, rather than an opportunity to speculate and gossip. It was as well that Bedford was present during these sessions, for McArthur was a man who took advantage of his position where women were concerned. On 18 March 1836, the editor of the *True Colonist* would accuse him of behaving in a way that would 'shut any man out from admittance into any well-regulated family circle', and he was soon to leave the colony in disgrace, pilloried as the 'kissing' parson.

These two men had much to accomplish before Mary McLauchlan was executed. At her trial, she had kept silent about many things, admitting no guilt and holding Mary Cameron's name close. Bedford, in particular, worked hard to obtain a full confession of her guilt over the weekend on which the Executive Council sat. It was his duty to reclaim her soul, but there were also other interests involved, for a person who went to the scaffold unrepentant, or worse, professing innocence and denouncing somebody else, brought forth a problematic response— from the point of view of the authorities—from those crowding to listen and watch. Public executions were meant to illustrate the majesty and power of the law and instil fear in the hearts of the audience, but in Hobart Town as elsewhere, many attended more for the spectacle of the thing.

On Sunday evening, after Hone's visit to Mary McLauchlan, when the Executive Council had finished sitting, and Jocelyn Thomas had lost the argument, Bedford finally succeeded in persuading her, as noted above, of two things. One required her silence. She must give up the idea of denouncing her 'seducer' from the scaffold. Instead, Bedford agreed she could name this man to the Sheriff and others who would gather in the prison's lobby before her execution on the following morning. Bedford's other demand involved a revelation from Mary McLauchlan. She must tell him the name of the woman who had helped her to murder the child. In his usual self-promoting way, Bedford shared his eventual success with newspaper editor James Ross, who wrote in the *Courier* (24 April 1830) about this final evening:

> The mental agony which this wretched woman is said to have endured previous to the execution, is described as truly horrible. Until 11 o'clock on the Sunday night, she only partially confessed her guilt to the Rev. Mr. Bedford, but when she had fully disburthened her mind at the time, and disclosed the person who assisted her in the murder, she became more calm, and towards morning, when the dread hour approached, she was resigned and penitent.

Bedford must have promised Mary McLauchlan that she would receive God's forgiveness in return for such penitence. He may have needed to bargain for it, offering her in exchange a promise to intervene to ensure that Mary Cameron would not hang, for the day after Mary McLauchlan's execution, her friend was discharged by Governor's Proclamation, and got exceedingly drunk.

Jocelyn Thomas was not the only person in Hobart Town who was uneasy about a capital sentence being handed down for this particular crime. To some extent, even the members of the Supreme Court must have shared in the unease, for court records show that during the following decades no woman in Van Diemen's Land was executed for murder when she killed her newly born infant; and the lesser charge of 'concealing the birth' of a child attracted a very light sentence: one year's imprisonment for Elizabeth King (1842) and Mary Ann Dempsey (1860); two years for Bridget Kenney (1854) and Isabella Thompson (1856).

The fact that the capital sentence was a contentious one was also illustrated in colonial newspapers. Rival presses provided a forum in which concerns about capital punishment were expressed. Ross's *Hobart Town Courier* was in favour of public executions. In general, Ross could be found on the day of any execution standing close to the scaffold in order to note carefully every feature of a convicted man's countenance and hear every word spoken from that powerful place. He wished to disseminate the moral lesson as broadly as possible through the pages of his newspaper.

In contrast, the anti-Arthur *Colonial Times* generally took a position against capital punishment, using public hangings as a different lesson, one about an authoritarian Governor who refused to countenance the extension of civil liberties to the free people who had settled in the colony. The *Times* and the *Tasmanian and Austral-Asiatic Review* (23 April 1830) expressed concern that a woman could lose her life for killing a bastard child while the well-to-do man who had fathered the child escaped punishment for impregnating and then abandoning her. In the case of Mary McLauchlan, rivalries fell into the background, for even James Ross worried about the fact that the colony was about to participate in the 'humiliating and … unheard of spectacle in this place of the execution of a female'. The *Tasmanian and Austral-Asiatic Review* pointed out that in England, though infanticide was 'not very unfrequent', in most cases juries returned 'a verdict for the minor offence, that of concealing pregnancy', which resulted in a term of imprisonment rather than a sentence of death. Despite his usual enthusiasm, Ross was so worried about this execution that he refused to attend it. Nevertheless, he still hoped that Mary McLauchlan's death would have a 'salutary effect', and wrote about the hanging ('this awful warning') in terms that provide us with as vivid a picture as if he had witnessed the proceedings in person:

> We have on several occasions made it a sort of duty to attend the execution at Hobart town of criminals of the other sex, because we hoped that a candid narrative of so terrible an example would not be without its influence, but we confess the present case was too painful for our feelings, and we were not present. Through the assiduous ministry of the Rev. Mr. Bedford, a certain hope of forgiveness supported her at the last hour; and she died contrite and resigned … When in health she must have been an interesting woman, with rather a pleasing countenance … How awful, and we trust impressive, the dreadful lesson thus held forth

Mary McLauchlan's Convict Record, 1828–1830 (AOT CON40/5). Perhaps Cook sensed that someone in authority was about to look over his shoulder and find he was illustrating the convict records. He has quickly pulled his pen away from this sketch, leaving a trail of ink on the paper. Courtesy of the Archives Office of Tasmania.

to both sexes in this peculiarly situated colony. Well has the first step to error been compared to the burning spark, which when once lighted, may carry destruction to inconceivable bounds.[35]

Then, a final annotation was made on Mary McLauchlan's convict record by Edward Cook, the convict clerk whose job it was to keep Governor Arthur's Black Books. He closed the record by writing, in unusually beautiful script, the words, 'Executed for the Murder of her Infant', followed by an incorrect date. Then he added something that was entirely his own, a rough sketch of a full-skirted woman hanging from a scaffold.

When Mary McLauchlan became entangled in the early nineteenth-century British system of crime and its punishment, she also became one subject for dissection about whom much can be learned, for records came to surround her, and they reveal much about the society in which these events took place. Exploring the relationships within which this woman lived, died and became a subject for dissection reveals something fresh about how British medical men were seeking to establish themselves in a new place. Two years later, the British Anatomy Act was passed, one clause of which rescinded the Murder Act under which people like Mary McLauchlan became the only legal subjects for dissection. But although Britain's Australian colonies had the ability to replicate such Acts, none did so at the time, and in Tasmania murderers continued to be dissected for many years.

CHAPTER THREE
INTERLUDE

ARY McLAUCHLAN HAS DISAPPEARED. ALL THAT IS left of her now lies in that unconsecrated earth in Hobart Town —except, that is, for any souvenirs that have been taken from the dissecting room. Those who knew her in Glasgow, including her two young daughters, will never learn what became of her in Van Diemen's Land. It is one of the many casual cruelties inflicted by the British government on those whom it sent into exile.

There was also, however, another context within which this episode can be read, for the year 1830 has a special resonance in the history of Van Diemen's Land. On that April weekend when Governor Arthur was thinking about the implications of executing a woman for the first time, he also had other important matters on his mind. Between those two meetings of his Executive Council, Arthur was crafting a letter to the British Secretary of State, in which he sought to justify the stringent measures he was about to set in place for dealing with the colony's Aboriginal people in the face of increasing violence. Like others, Arthur believed the original inhabitants of the island were at risk of extinction in the face of British settler-colonialism.

Following his thoughts in this direction, it is time to let Mary McLauchlan go and look at some of the other people whose bodies attracted the curious gaze of nineteenth-century medical men. Anatomy was a comparative science. The men practising it compared normal and pathological, female and male bodies, and looked for anatomical similarities and differences between what they took to be the different races of humankind, and between human and non-human beings. Vast collections of bodies and body parts were necessary for this science, and medical men were major contributors to such collections. In Van Diemen's Land, soon to become Tasmania, they obtained Aboriginal bodies by disinterring them from their graves and dissecting those that came into their hands in the colony's hospitals. And due to their comparative rarity, Tasmanian Aboriginal bodies quickly came to hold a special place in British and other European collections.

At first glance, it might seem that this move from gender to race, and from Van Diemen's Land back to Britain, is a clumsy one. Yet, in Van Diemen's Land at this time, people experienced their lives in ways in which gender and race were interrelated, as were the punishments for crime and the apparent disappearance of the colony's Aboriginal people. Some say the workings of these cultural phenomena continue to stain the Tasmanian psyche. Both, too, were a matter of bodies and the everyday practices through which some people were turned into things by others. Medical men and their work on the dead remain our focus, though the dead upon whom they went to work now changes.

The racial aspect of Mary McLauchlan's life in Van Diemen's Land was not immediately evident to me. But while searching everywhere in contemporary records for a sign that someone else was reflecting on her death, I kept stumbling across references to 'the Tasmanians', as the Aboriginal people of that island were then called. The men who had been arguing that Mary McLauchlan's execution was a sad turning point in the colony's history seem to have forgotten her almost immediately after the execution. Instead, everyone was talking about the Tasmanians, whom it was becoming increasingly difficult to ignore in the face of murder and mayhem in Hobart Town's outlying districts.

James Scott, William Bedford and Jocelyn Thomas were busy meeting with others as members of Governor Arthur's Aborigines Committee. They had been set the task of gathering settlers' opinions about the cause of the Tasmanians' increasingly violent response to the presence of Europeans everywhere in their land. Men on the committee gathered no opinions from the Tasmanians, and none from the assigned servants, who were most commonly the victims of these murders, though they had made no choice to 'settle' in Van Diemen's Land.

Mary McLauchlan's death, we now know, was part of a much larger tragedy. It is not necessary to narrate here everything that happened during cultural contact in Tasmania; that history has been and is being written by others. The facts are bald and frightening, no matter which side you take in the current acrimonious debates about Australia's past. When the British arrived in Van Diemen's Land in 1803 from Sydney Cove, several thousand people already inhabited the island. They were the descendants of others who had lived there for perhaps 60,000 years (though that was not known in 1803). Yet by the time Charles Darwin arrived in Hobart Town on the *Beagle* for a brief visit in 1836, all he noticed about the Tasmanians was their apparent absence. Only some two hundred remained alive, and these people had been persuaded by George Augustus Robinson, a man of missionary zeal, to remove themselves to Flinders Island, to the north of Van Diemen's Land, in Bass Strait, for their own good.

Forty years later, all of the people whom the Europeans recognised as 'Tasmanians'—which meant only those of full Aboriginal descent—had died from European diseases to which they had no immunity, from murder, from a low fertility rate, and from desperation at being exiled from their own land. As a race, they were considered to be extinct, in what was expected to be the first among many such disappearances in the face of contact in all parts of the globe between European and native peoples. Some thought it a sad but inevitable outcome of the failure of an inferior race to thrive in the presence of a civilised people; in Patrick Wolfe's words, in the presence of 'their immeasurably distant future'.[1]

Over the years, successive colonial Governors and British Secretaries of State for the Colonies expressed the hope that in Van Diemen's Land it would be possible for 'natives' and Europeans to coexist. British colonialism was meant to be quintessentially different from that of other European powers. Given contemporary understandings of the place of this penal settlement in the British world, that hope, if sincerely held, seems to have been extraordinarily optimistic. On this island, cultural contact was a meeting between people believed to be savages and

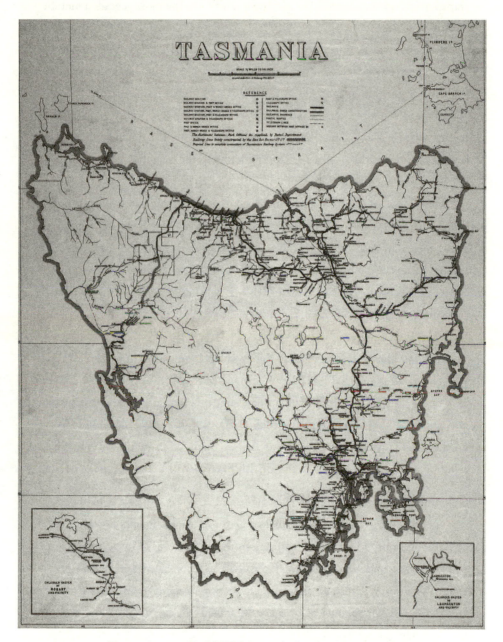

Tasmania, c. 1890 (AOT Map 18), showing Hobart and Oyster Cove and—to the north-east of the main island— Flinders Island. Courtesy of the Archives Office of Tasmania.

transported felons understood in their own land to be the most degraded members of British society.

When George Arthur arrived to govern the colony in 1824, he brought with him progressive credentials where race was concerned. Arthur was an abolitionist who, in his previous assignment as Governor of Honduras, had earned the anger of European sugar planters by insisting they had no right to inflict injuries on the bodies of their slaves. This story had preceded him to Van Diemen's Land, and several of the wealthiest settlers drafted a letter to the Secretary of State requesting he not be appointed.

Arthur was among those who believed that the Tasmanians were a race of human beings capable of 'improvement', with the helping hand of civilised and godly men. He disagreed with those who designated them 'the lowest order of human beings, removed but one shade from brutality'. Instead, he thought they were comparable to the 'hundreds of Irish, English and Scotch [who] are, at the present day, as wild and as ignorant as they'. For Arthur, the Tasmanians possessed 'a considerable amount of latent capacity'.[2]

However, the process of tapping this capacity for civilisation was greatly complicated on an island that had been incorporated into the British colonial system as a penal settlement, one in which increasing numbers of settlers were arriving and taking land. The Governor of such a colony was necessarily simultaneously engaged in many different projects, as Arthur pointed out in 1828 in some exasperation to William Huskisson, the British Secretary of State. He spoke of the

> weight of responsibility incident to a new, rising, distant, and most perplexing Government—with Lands to grant—Emigrants to settle—Institutions to form—the Instructions of His Majesty's Government to observe—and the colonists, if possible, to conciliate—(every one of which duties involves a variety of conflicting interests)—and, above all, with some thousands of depraved Convicts to assign, discipline and employ ...[3]

Managing settler-colonialism was a particularly complicated undertaking in this place. It was a system premised, as Wolfe has argued, on taking the land and eliminating its native societies.[4] But this did not mean eliminating all individual 'natives', whom many believed could be 'civilised' and take their place among those who had replaced them on the land.

Four years after Arthur's arrival on the island, he was faced with a settler outcry about escalating acts of violence by the Tasmanians. In one month during 1828, in the district of Oatlands, twenty-two inquests were held into the deaths of Europeans who had been murdered. To Arthur the need to take some decisive step was becoming more apparent every day. He established roving parties to capture the Tasmanians, instructing them to use arms only when all other measures failed. In April 1828 he banned Tasmanians from entering the settled districts and, in November of that year, he took the extraordinary measure of declaring martial law. The violence continued on both sides.

By the weekend when Arthur's mind was said to be occupied with determining Mary McLauchlan's fate, the Governor was also busy writing an important letter to the Secretary of State, Sir George Murray, to prepare him for the news that further extraordinary measures were required. Arthur informed Murray that different opinions existed in the colony about how to calm the crisis. He said, 'many respectable inhabitants in Hobart earnestly desir[e] a continuance of conciliating measures, while the residents in the interior deemed more severe measures essential to their preservation'. With this letter, Arthur enclosed the report of his Aborigines Committee, which was prefaced with a regret that in Van Diemen's Land, the English had experienced a 'forgetfulness of those rights of ordinary compassion to which, as human beings, and as the original occupants of the soil, these defenceless and ignorant people were justly entitled'. The report then went on to speak of first causes where the violence was concerned.[5]

In establishing the committee and asking its members to enquire into the historical roots of contemporary violence, Arthur was making history. Most of the settlers who expressed their opinions to the committee placed the blame for the violence with certain debased Englishmen. Individual miscreants were named, men who were said to be 'a disgrace to our name and nation, and even to human nature'. They had turned the 'naturally mild and inoffensive' Tasmanians into violent men. The Tasmanian response was understandable: 'A worm ... will turn'.

One man, known as 'Carrots', served an emblematic function in this history-making. He was an escaped convict who boasted of decapitating a Tasmanian man, then stealing his wife and making her wear her husband's head tied around her neck 'like a plaything'. It was probably a hideous parody of Tasmanian funerary practices. Carrots illustrated the depths to which an Englishman could sink.

In his letter to Murray, Arthur emphasised that such men were beyond government control. Those who treated the Tasmanians so cruelly were 'lawless convicts' who had 'from time to time absconded'; 'distant convict stock-keepers in the interior'; and sealers employed 'in the remotest parts of the coast'.

Gilbert Robertson, the Chief Constable in Richmond (near where Mary McLauchlan had served the Nairnes), provided evidence to the committee of a similarly depraved individual on the Tasmanian side. This was Musquito, a Sydney 'black' who had been transported to Van Diemen's Land in 1813 for murdering a woman. Robertson named Musquito as the cause of the increase in Tasmanian violence after 1824, when he began to lead the Oyster Bay people in attacks against the settlers. Musquito had corrupted a previously benign group, turning them into cruel men who perpetrated Carrots-like acts of barbarity on the bodies of those they killed.

Through Carrots and Musquito, this violence became a matter of convicts and natives, in a conversation that took for granted people's shared understanding of such shorthand ways of speaking. It turned the Tasmanians' violence against settler-colonists into crimes of the sort the British were used to. The rest of Vandemonian society could think itself different and search for a rational solution to the problem.

The men on the Aborigines Committee thought the wisest policy would be one based in justice. They acknowledged that the British had taken the Tasmanians' land, and recommended to London that all future attempts at colonisation should be grounded in an understanding that a strict 'obligation exists to exercise mercy and justice towards the unprotected savage'. Otherwise, retribution would follow. They spoke as 'men, as Englishmen, and as Christians', and called for moderation and forebearance 'to an ignorant, debased and unreflecting race'. However, it is difficult to understand how forebearance underpinned some of the committee's recommendations, which included a suggestion that force be used to resist force by arming settlers so they could protect what had so quickly become their own property, and by attaching mounted police to each station.

Arthur also received many other suggestions from settlers. One proposed building 'a strong compact hut for six missionaries who would gradually and imperceptibly create friendship with the blacks'. Another recommended capturing the Tasmanians' leaders and conciliating their followers. A third suggested pre-emptive strikes against the natives 'in their recesses in the woods'. A fourth believed the natives should be removed from the island lest they be 'hunted down like Wild Beasts and destroyed'. There were also plans for enticing the Tasmanians into purpose-built huts containing secret rooms, trap-doors and spring-locks; importing black-trackers from Sydney; training dogs to hunt the natives down; and inviting 'New Zealand cannibal chiefs to bring over some of their tribesmen to hunt down, enslave, or otherwise get rid of the Vandiemenslanders ... and promote good [trade] relations with New Zealand and, maybe, help to civilize the Maoris'.[6]

A few months after he forwarded the committee's report to England, Arthur —who was, after all, a military man—decided to act in a military way. He devised a campaign that combined military men and settler volunteers to form a 'Black Line' across the island. Moving gradually south, these men were instructed to drive the Tasmanians before them onto a peninsula on which they could henceforth live, kept safely apart from the Europeans. At least that was the plan. It was a spectacular failure. Settlers hostile to Arthur said it was 'as sensible as trying to harpoon whales from the heights of Mt. Wellington'.[7] Only one old man and a boy were captured during the campaign, and they soon escaped.

Those non-military men who had volunteered to take part in the operation were hopelessly out of their depth in this land. What looked like a straightforward operation on a map was a very difficult thing to accomplish in an unfamiliar landscape made up of mountains, narrow valleys and much territory that the British had never visited. One eager volunteer, Henry Emmett, a clerk in the office of the Principal Superintendent of Convicts, has left a record of his participation in the Black Line. He did not mean it to be read as farce, though all the ingredients for comedy are there. After excitedly kitting himself out in a specially designed moleskin suit, Emmett gathered his men to listen to Arthur's stirring speech before marching them off into the interior, where he experienced problems from the beginning.

At first, the men kept disappearing into the public houses they passed along the route and returning drunk. Further out from Hobart Town, Emmett found that the settlers whose properties they passed—whose properties they were working to protect—resented supplying them with food. His moleskin suit became soaked with rain and never dried out; the campsites he chose turned out to be swamps or, once, to contain a nest of snakes; and the only signs of Tasmanians he ever saw were their distant fires.[8]

By 26 November 1830 the operation, which had cost £30,000, was over, and Arthur was searching for another solution. Enter George Augustus Robinson, a carpenter and evangelical Christian, who offered to travel into the hinterland in the company of a group of friendly natives and persuade the rest of the Tasmanians into voluntary exile.

Histories have been written about the unexpected success of Robinson's 'friendly missions', which took place over five years from 1830. The missions resulted in 201 Tasmanians being 'brought in'. From 1833 to 1847 these people resided on Flinders Island, where they were subjected to the benefits of civilisation and religion. But the island became a place of death. In 1838, when Robinson left Van Diemen's Land to become the Chief Protector of Aborigines at Port Phillip (Victoria), he left behind just eighty-two Tasmanians, of whom only fourteen were children. Nine years later, when the Tasmanians were moved back to the mainland and settled at Oyster Cove, only forty-seven remained alive.

Their old 'conciliator' visited them there in 1851. Interestingly, he spent time making a careful sketch of where those who had died following the move were buried. It seems odd behaviour, until you remember that this man had no compunctions about collecting Aboriginal skulls. When he returned to England, in 1852, he carried several with him, having already made a gift of one to Lady Jane Franklin, wife of the Governor who succeeded George Arthur in Van Diemen's Land.

By the mid-1850s, the British were making contributions to the European scientific narratives about the Tasmanians, whose extinction seemed imminent. This turned scientific work on them into an urgent project in which colonial men contributed by collecting the material that informed European debates. They gathered the Tasmanians' words to make dictionaries, and samples of their material culture in the form of tools, weapons, baskets and necklaces made from shells. They collected information about their burial and mourning rites. And, in the knowledge that the Tasmanians held strong beliefs about the due treatment of their dead, medical men contributed in the way to which they were ideally suited. They harvested their bodies.

Many of these body parts were sent to Europe, where metropolitan scientific gentlemen studied and interpreted them. Thus links were formed between colonial and metropolitan men through the exchange of Aboriginal bodies for rewards, and our focus on crime and its punishment as a way of gaining access to the dead is being replaced by one revealing scientific work on race and its … delights.

Plan of Oyster Cove Station and Graves, 1851, by George Robinson (ML, G. A. Robinson Papers, Vol. 67, A 7088, p. 47). Robinson's sketch reveals where Aboriginal and non-Aboriginal people were buried, though only the Aboriginal graves are named. Robinson was not even sure how many Europeans had been buried here ('about 10 whites'). Perhaps their graves are only marked as a warning where not to dig when seeking valuable bones. Courtesy of the Mitchell Library, State Library of New South Wales.

CHAPTER FOUR
THE BONE COLLECTORS

IN BRITAIN, COLLECTING HUMAN BONES WAS A KIND OF MID-Victorian mania, shared by amateurs and professionals alike. All felt that they were contributing to the study of British antiquity. Where human skulls more generally were concerned, Dr Joseph Barnard Davis was the epitome of collectors. By 1867 he had amassed 1474 of them, which made his collection larger than those of all the public museums in Britain put together—in fact, the largest in the world. All crowded into the house of a country doctor, in the market town of Shelton in Staffordshire. This must have been disconcerting for the patients who came to consult him. It certainly frustrated his fellow anthropologists, who could not understand why such a 'superb collection' should be housed in such an 'out-of-the-way' place.[1]

Like many others engaged in this kind of collecting, Davis was a medical man. When he died, William Henry Flower spoke of Davis as leading 'the simple life of a medical practitioner'.[2] Clearly, in mid-nineteenth-century England, such a life could encompass much more than caring for the health of patients. Exactly how much more is revealed in Davis's notebooks and anthropological publications. Here was a man on a scientific quest to untangle the puzzle of the peopling of the world. He set about making the contribution of an avowedly practical man to speculations about human origins and diversity, collecting the kind of raw material upon which this science was built.

Davis's transformation from medical man to physical anthropologist can be traced in a series of notebooks in which he began writing soon after qualifying for medical practice in 1823. The opening pages of the first book contain neat notes of medical conditions that interested him. However, soon the bulk of his writing comprised meticulous records about the appearances of the people with whom he came into contact. He was beginning to observe everyone with new eyes and to categorise them into racial varieties.

Joseph Barnard Davis, date unknown, by R. J. Lane.
Courtesy of the Wellcome Library, London.

Writing in 1867, Davis traced his growing interest in such matters to his medical training (and an intriguingly unexplored 'accidental conversation with a friend').[3] Since he had learned anatomy at Joshua Brookes's private anatomy school in London, this is not surprising, for Brookes emphasised the importance of comparative anatomy and offered his students lavish opportunities to dissect—to the obvious chagrin of the men teaching in London's large charitable hospitals, who competed against him for the custom of fee-paying students.

Their resentment showed in contempt for Brookes. The Fellows of the Royal College of Surgeons sneered at his dirty hands and dishevelled appearance which, to them, marked him as something other than a gentleman. He was not among those invited to view the official dissections of murderers' bodies at the College. John Flint South recalled Brookes as the 'dirtiest professional person' he had ever met, although he admired his excellent school. Eventually, the College Fellows managed to make Brookes's school unprofitable by refusing to recognise its students' qualifications.

Under Brookes's tutelage, Davis became a skilled observer of anatomical differences, and it may also have been there that his interest in collecting was encouraged. When Brookes's museum of comparative anatomy was later sold, his former student purchased some of the skulls it contained.

An early and roughly handwritten catalogue of his own skeletal collection appears among Davis's medical notes. At this stage his interests were eclectic. He possessed several skulls, including one of a polar bear ('killed in the Greenland Seas in 1820, and which I brought from thence') and another of an old man, with lower jaw attached, which he had purchased at the sale of a surgeon's effects in Derby. These items joined a Roman skull, found by a lady in a cellar in the Suburban Villa of Diomedes at Pompei; and the extremely long skull of an old woman who had died in a prison in France. Davis had this particular skull read by a phrenologist, who informed him the woman exhibited one of the worst of dispositions. He also collected the skulls of nine Chinese pirates who were executed at Ningpo in 1852.

This collection indicates the apparent ease with which metropolitan scientists obtained the bodies of those who died in far-flung lands. No less than *nine* Chinese pirates—as if they could be purchased in job lots.

The fact that Davis was a medical man was important. Making fine distinctions between normal and pathological body parts was a matter of proper training for such men. This discerning ability was often joined with a particular, nineteenth-century way of seeing that facilitated the making of a cartography of human differences. For medical men, anatomy served several purposes. Robert Knox, both a philosophical and a practical anatomist, articulated it in this way:

> Anatomy is not a science, but merely a mechanical art, a means towards an end. It is pursued by the physician and surgeon for the detection of disease, and the performance of operations; by both to discover the functions of the organs; and by the philosopher with the hope of detecting the laws of organic life, the origin of living beings, and the transcendental laws regulating the living world in time and space.[4]

GATHERING FACTS FOR THE SCIENCE OF MAN

Nobody was safe from Davis's speculative gaze. Wherever he went, he carried a pocket-sized book in which to jot the observations he made about people, and these notes became the base material from which he later wrote a more polished account of regional differences between the British (published in 1865 as *Crania Britannica*).

Davis's sometime collaborator, John Beddoe, had developed a more discrete method. He credited the small recording card he held in the palm of his hand with the fact that he seldom met with any real difficulty on the part of his subjects. He thought another factor was his ability to merge into the crowds at the local fairs and markets, which were his best 'hunting-grounds'.[5]

The word pictures contained in Davis's notebooks make fascinating reading. In them, we catch a fleeting glimpse of a 'Highland Woman' dancing to music in the street and 'playing cassanets [*sic*]'. It was not her dancing that attracted Davis, but her height—she was 5 feet 8 inches tall and, when questioned, told him that her younger sister was even taller. Sarah Jones, 'a Gypsey woman … of the usual tawney colour', has also been captured in the notebook, as has William Miller, ten months old and spied at the railway station in Stafford. He was worthy of note because of his 'very curious forehead'.

Davis sometimes puzzled over his observations because they clashed with his expectations. Passing a 'healthy florid young girl' with grey eyes, he could not understand why her hair was 'white, with a very slight tinge of red'. The appearance of two young women, both from Galway, differed so sharply that he could make no regional sense of them at all. Such differences posed a problem for a project that sought to pin down group-based, rather than individual characteristics. Nevertheless, Davis became increasingly certain of his own evaluations. When a man—questioned as to his nationality—replied that he was a Spaniard, Davis wrote down in his notebook 'a so called Spaniard from Bengal'. After probing further, he noted that the man 'did not deny that there was some mixture in his blood'. And when a military gentleman claimed to be Hungarian, Davis added, in brackets, 'a Jew of Amsterdam I think'.[6]

The doctor's patients joined these strangers in his notebooks, with an extraordinary degree of insensitivity on Davis's part. In other respects he was not an unfeeling man, so perhaps he understood his probing as necessary for a scientific project that was a matter of some urgency. When a tailor named 'Jaun de Bure' came to Davis to register the death of his child, the doctor seems to have spent most of his time making notes about the man's physical appearance and descent. In response to a question from Davis, de Bure said that he was French, though Davis thought that the man's features 'strongly marked' him as German. 'Of course,' he noted, I 'could not contradict him. I put his name down Jean de Bure, and afterwards put the pen in his hand to sign his name, when he wrote Jaun de Bure [which] revealed to me that he was not of French but German blood'. This is all that is left to us of de Bure, who must have wondered why Davis seemed to care more about the purity of his lineage than the death of his child.

Davis undertook this observational work in the knowledge that in Britain the regional differences he was seeking to capture and pin down were melting before his eyes. Rapid industrialisation was transforming the demographic landscape as people migrated in search of work and formed sexual relationships across regions. The result was a hybridised population. In April 1860 Davis made an urgent note to himself: 'Visit rural parishes and smaller manufacturing villages … to see what change manufactures have produced'.[7] He also wrote to solicit 'Friends and other Gentlemen' to join him in capturing the peculiar features of lineage and descent among those who lived in rural and remote areas, before any further amalgamations occurred.

The gentlemen Davis addressed were those placed in 'situations favourable for ascertaining the physical and other peculiar characters of the people surrounding them'. This was a job description perfectly suited to medical men. Davis solicited 'a few facts' to help in scientific investigations into matters bearing on 'national character'. He sent the gentlemen a set of instructions to guide their observations, and asked each to seek 'at least *twenty adult males* of average character'. They were to notice such things as stature, bulk, the 'character' of faces, the colour and texture of hair and the colour of eyes.

Davis also required them to note the size and form of the men's skulls. These observations, he instructed them, could be 'easily ascertained by passing a tape, graduated in inches and 10[ths], round the head at its greatest circumference'.

As well as assessing the living, Davis instructed his gentlemen to obtain the skulls of the dead (both ancient and modern) wherever possible and also, if available, photographs, prints, or drawings. In addition they were to elicit family names (both common and 'peculiar') and assess the race of the inhabitants in the area, together with its relative purity: 'Do the inhabitants often marry strangers, or have they kept their blood pure?'[8]

The collation of this material added to Davis's storehouse of facts. In particular, he knew that bones were reliable objects and viewed his work with British skulls as complementing that undertaken in America by Samuel Morton. Morton possessed the largest collection of what a visiting scientist called, in 1851, 'craniological treasures'.[9] They were catalogued in his *Crania Americana* (1839) and *Crania Aegyptiaca* (1844). The collection comprised 1035 crania, which he had divided into categories of Race, Family, Tribe, and so on. Lying among them was the skull of the Vandemonian cannibal, Alexander Pearce, which James Scott had allowed to travel from his Hobart Town dissecting room, via Calcutta, to Morton's museum. It sat on a shelf there, somewhat oddly labelled 'ENGLISH' and 'cannibal', and situated between the skull of a modern-day Italian mass murderer and that of an ancient Egyptian mummy. Morton poured lead shot into each of his skulls to measure their cranial capacity and so enable him to rank the different races of humankind. Davis admired this 'magnificent' achievement, and was adamant that Britain should not be left behind in the illustration of national character.

The doctor collected some skulls personally. In April 1859 he and William Durlyneux gained the Duke of Sutherland's permission to open two ancient British burial sites in Trentham, Staffordshire, to see what they contained. The collectors were fortunate in obtaining the assistance of a certain Mr Swynfer Jarvis, whom Durlyneux had recommended for the job after finding him capable of 'digging with a vengeance'.[10]

They found a quantity of bones that had been burned. In 'Various Notes', Davis wrote that it 'was difficult to avoid the dreadful surmise that the Chieftain of his Tribe … had been accompanied to the future world by a wife or slave, immolated on the spot, with the vain purpose of being his attendant in the Elysian fields. The excavations of British barrows reveals satisfactory evidence that the funereal rites of the Aboriginal Tribes were observed by such cruel practices.' Davis was a man with little time for religion, pagan or Christian. Some judged him harshly for the vigour with which he went about his bone collecting.

He sought to pass on his own skill to others engaged in disinterring bodies from barrows in Britain at the time. His colleague John Beddoe commented that Davis's 'enthusiasm for his subject was wonderful, but sometimes it verged on the ghoulish', for Davis looked at the heads of the living 'simply as potential skulls'.[11]

He always strove to ensure that those engaged in all this digging on his behalf understood that they must collect skulls in a way that would make them of use for the purposes of science. In October 1853 he wrote some 'Hints for Collecting and Preserving the Bones of Ancient Skulls' and sent it off for publication in the *Gentleman's Magazine*. Its tone was more like a set of instructions than hints. Scientifically minded diggers, Davis asserted, must be sure to collect 'the whole of the bones of the head and face … not a few fragments, or the mere brain-case'. They were to be especially careful to preserve the slender bones of the nose, which were thought to be 'very characteristic'. When they found a skeleton, they should proceed carefully from the feet upwards, laying aside pick-axe and shovel and employing instead a piece of wire to free the skull from the earth and gently raise it. The skull was then to be labelled with the name of the barrow and a number, and the same number was to be inscribed 'with a pen upon the skull', which should then be placed on a piece of soft paper and immediately wrapped into a package which was to be tied with string. Nothing was left to chance in these instructions, and Davis turned to poetry to express the importance of tender care:

> All, all have felt Times mighty wand
> And, brought again to light
> Defaced, despoil'd, can scarce withstand
> The touch, however slight.

Finally, in these neat parcels, the skulls were to be packed in a box with a little hay for safe transport to his collection.[12]

The skulls gathered in such ways were understood to illustrate racial differences within the British Isles. However, Davis's project was global in scope, and he also collected facts that would illustrate greater differences. He was a relentless scavenger of bones, intent on making the kind of collection that would illustrate such sharp divergences between the races that no-one could deny they were akin to different species. This goal rather put paid to his continual insistence that he was a practical man who was merely engaged in the task of collecting facts upon which *others* could build such systems of thought.

Davis was one of many Victorian men trying to determine the answers to fundamental questions about humankind, in light of increasing understanding that the earth's history was exceedingly deep. Prior to the publication of Charles Darwin's *The Origin of Species* in 1859, there were two main threads to this scientific debate: the monogenists believed humankind had a common point of origin and that differences between the races were a matter of degree rather than kind; the polygenists held that the races were both distinct and immutable, and had different origins. For polygenists, there was no way that civilised men with good intentions could 'raise' the savage races. They occupied their time in searching for characteristic 'types' of each species and mapping the differences.

As George Stocking notes, the result of this search for characteristic types was that a

> fictive individual who embodied all the characteristics of the 'pure type' grew in the imagination, obliterating the individual variation of his fellows, until he stood forth for them all as the living expression of the lost, but now recaptured, essence of racial purity.[13]

Polygenists were also preoccupied in thinking about human reproduction. They believed that racial immutability could be proven by illustrating that sexual unions between civilised men and savage women, like those between horses and donkeys, produced no fertile progeny. Although German anthropologist Johann Blumenbach had argued against this in the late eighteenth century, the theory still held weight. Men such as French anatomist Paul Broca, who founded the Anthropological Society of Paris in 1859, developed complex scales and glossaries of terms in which this view was given scientific credibility, and Tasmania played a part in his thinking.

Broca knew that on this island a large number of British men lived alongside a population of 'savage' women. He thought it made the colony a place ripe with interracial sexual possibilities, yet few children had been born from unions between British men and Tasmanian women. Could this be due, he wondered, to the 'ugliness and dirty habits' of the native women of Australia? Had these 'bridled the sexual desire' of European men? Broca dismissed the thought, for he knew that no seaport contained a 'prostitute sufficiently ugly and old to frighten the sailor'. Therefore, something else must be at work. So he placed the 'mongrel' progeny of unions between European men and Tasmanian women in his 'oxygeneric' group, which made them a people who would die out within two generations.[14]

Despite their different views, monogenists and polygenists shared some priorities. For both, distinctive races of humankind did exist empirically, and the differences between them were so marked that they needed to be explained and could be mapped on a linear scale of humankind, with savages like the Tasmanians at one end and Europeans at the other. Much of the disagreement revolved around exactly *where* each race should be slotted and what provided the best measure of difference between them. This is where the facts provided by practical men came into their own. As Paul Turnbull has shown, skulls came to occupy a special place in the empirical investigation of human differences. For men such as Blumenbach, they were the kind of substance upon which a history of human diversity could be plotted.[15]

In his quest to find characteristic skulls, Davis forged fruitful connections across the British Empire, as well as in America and Europe. He undertook this work in the strong belief that it must precede high-flown theoretical understandings of race. The resulting *Thesaurus Craniorum* was, he stated, 'a mere Catalogue' that made no attempt to 'develop or to support any great theory', though he was aware that this would disappoint 'some excellent persons, who look upon nature as of value only as so much material to be worked up into hypotheses'.[16]

Later, when writing up the results of his examination of Tasmanian skeletal material, he took the opportunity to expand on this preference for gathering facts. He wrote, '[systems] are generally built up by an incorporation of much that is imaginary ... [and] are often unable to withstand an appeal to facts'.[17]

CRANIAL FACTS

Davis's skull collection was viewed by many as a veritable treasure-house. He gathered the material that was available to him as a British man. This meant that he had no need to travel far to obtain exotic specimens, for he could forge links to Englishmen abroad. Being at the centre of the empire, he also had access to living specimens of exotic differences that were brought to England to be observed first hand. During the 1850s, as a member of the Ethnological Society, Davis would have attended meetings at which members were provided with opportunities to examine 'living aboriginals' from far-flung lands. He may also have attended Exeter Hall in London in 1847 to see Robert Knox's display of a group of 'sullen, silent, and savage' Bosjesmans who had been brought to England by a Liverpool merchant.[18]

In addition, Davis provided himself with opportunities to inspect those belonging to mysterious races. It was a time in which strange sightings seem to have occurred. The *Anthropological Review*, published by the Anthropological Society of London, which prided itself on making anthropology into a hard science of facts, contained the following snippet in 1867:

> A man half white and half black has arrived in New York from Arkansas. One entire side of his body is almost black as ebony, while the other side is of the pure Caucasian hue. There is no humbug about the man. He seems very intelligent, and is desirous of avoiding public observation.[19]

In 1859 Davis met Ira Aldridge, the 'African Tragedian' and progeny of a 'native African-negro & … a Creole mother'. This gave Aldridge a special kind of factual status in light of debates about the children of cross-racial couplings. Davis noted the man's physical characteristics and also his 'manner, voice, mode of expression and style of conversation'. They had all the attributes of that 'which is called gentlemanly'. The two men conversed, though it is easy to see which of them chose the topic of their conversation. Aldridge gave Davis his opinion on the 'mulatto' question. He had not, he said, observed that people of mixed blood were 'feeble or sickly', though he understood that they were usually 'short-lived'.[20]

Six years later, Davis found another rare prize, this time in the street. He thought he had stumbled across 'a Hottentot', standing with his back to a wall, begging. But when questioned, this man informed Davis that he was a South Australian, which was authenticated for Davis by the fact that the man could show him a peculiarly painted cloth he had brought with him to England. The man's colour, Davis noted, had misled him.

There were other methods for observing racial differences without leaving Europe. Some were provided by popular anatomical museums, such as that belonging to Dr Kahn in London, which displayed specimens of comparative anatomy to a paying public. When medical student Shepherd Taylor visited it in December 1860, he thought it a 'decidedly indecent pseudo-scientific affair' and a 'sink of iniquity', though he returned the following day.[21]

Other opportunities were presented when people, such as a 'Hottentot Venus', died in Europe and their bodies fell into the eagerly waiting hands of anatomists.

William Flower spoke admiringly of the system that operated in Paris, where the School of Anthropology contained laboratories

> in which … all the bodies of persons of outlandish nationalities dying in any of the hospitals of Paris are dissected by competent and zealous observers, who carefully record every peculiarity of structure discovered, and are thus laying the foundation for an exhaustive and trustworthy collection of materials for the comparative anatomy of the races of man.[22]

In 1865 a young man became the subject matter of a scientific paper written by Dr Jeffries Wyman on the anatomy of the Hottentot. Wyman opened with a sentence containing his only acknowledgement that the object beneath his knife had been a person: 'The subject', he wrote, 'was nearly adult, and came to his death by suicide'.[23] Under what sad circumstances, we can only imagine.

Other foreign people travelled to Europe post mortem. Davis worked diligently to obtain such remains, making many purchases through the market in human bones. His papers contain catalogues and newspaper clippings of the sales he attended, and he also contemplated buying Dr Rogers's collection, which contained about 200 crania including, Davis thought, 'about 4 Tasmanians', as well as crania from the aboriginal peoples of North America, New Zealand and mainland Australia, and Cingalese and Esquimaux. Davis estimated the value of this collection at £100, and was especially pleased to find that the price included Dr Rogers's 'little green book', which contained a catalogue that gave each skull a reliable provenance.[24]

In addition, Davis pursued such specimens via more direct avenues, corresponding with travellers, collectors and residents in foreign lands. As always, he sought to ensure that the local digging these men undertook on his behalf was done in a way that would make the results of benefit to science. Inside one of his exercise books lie some loose sheets headed: 'Notes on the preparation of Crania in hot climates, and chiefly applicable to India'. The notes related both to animal and human material, and in them he talked collectors through the task of disengaging bone from flesh. They were to begin by removing the 'soft parts from the head whilst fresh', before macerating the crania in a large body of cold water, a beer barrel with a drain being ideal for this purpose. All red particles of the blood were to be discharged in this way. For human crania, Davis's 'Notes' instruct the collectors to

> break down and remove the whole of the brain through the foramen
> magnum before beginning the maceration, both to diminish the intoler-
> able fetor, and also to facilitate the discharge of all the blood by steep-
> ing, since upon this part of the process essentially depend the beauty
> and whiteness of the preparation.

Teeth were also a matter of importance in Davis's view of the world, for their size could establish with some precision how primitive the race to which a man belonged had been. As Darwin's ally Thomas Huxley noted, teeth were also indicative for evolutionists, who believed they illustrated the superiority of white men 'in [any] contest which is to be carried on by thoughts and not by bites'.[25] Davis instructed his collectors to preserve the teeth by detaching the lower jaw. If any fell out in the process, they were to be fastened back in 'with a little cotton and thick mucilage of gum arabic'. He felt sure that with such care, almost every cranium could be prepared.

COLLECTING TASMANIANS

In the political economy of bone collecting, Tasmanian skeletal material was highly prized. British medical men had been gathering Tasmanian bones since 1804, when several Tasmanians were shot at Risdon Cove. The penal settlement's surgeon, Jacob Mountgarrett, gathered up one body, which he preserved and sent in a barrel to Port Jackson.[26] By mid-century there was some urgency to the quest, for it was believed the Tasmanians would soon be extinct. Davis learned all he could about them through books and by making visits to museums, and he set about adding Tasmanian bones to his collection by developing links with settler-colonists who were in a position to obtain such material for him. He made the acquaintance of Josiah Spode, ex-Principal Superintendent of Convicts in Van Diemen's Land and the man who had measured Mary McLauchlan for Governor Arthur's records upon her arrival in Hobart Town. After returning home to

England, Spode lived near Davis, and put him in touch with Alfred Bock, whose father Thomas had made those post-mortem sketches for James Scott during the 1830s. Bock had also painted portraits of the Tasmanians.

In 1856 Davis wrote to Alfred Bock and spoke of his desire for Tasmanian bones. As was his custom, he set out with some precision the manner in which colonial men should obtain them on his behalf. It was a method designed to avoid being detected and thus giving offence. He told Bock that he himself could easily 'abstract skulls from dead bodies without defacing them at all, and could teach any medical gentleman to do this'. It was clearly a well-honed technique and required private access to at least two dead bodies, one from which to abstract the desired skull and another whose skull could be inserted in its place to disguise the theft. In addition, he suggested that a cemetery raid might be profitable.

Foreseeing Bock might object to this idea, Davis reminded him firmly that there are always difficulties, but these *may always be overcome*. He asked him to find a medical gentleman connected with public hospitals who would be willing to help him acquire Tasmanian skulls.[27] How well he understood who would be most able to assist him in his quest.

There is no evidence that Alfred Bock acted upon these suggestions, though someone else did. The letters exchanged between such colonial men and the English scientists on whose behalf they set to work illustrate how relationships between peripheral and metropolitan men of science worked. These men have respectively been termed foragers and cultivators of science. The letters reveal some of the complexities involved. Marcel Mauss's anthropology on the political economy of gifts further illuminates the workings of such relationships.[28] Through gifts, people are bound into reciprocal relationships with each other, in a culture of exchange that comprises giving and receiving. Invitations of friendship and social intercourse are offered via gifts, obligations are accepted or refused, and these exchanges are permeated with notions of courtesy, honour and prestige.

Tasmanian bones came to be the material of such relationships of exchange. They were harvested from flesh and disinterred from graves to become culturally laden gifts. Through them men such as Colonial Surgeon William Lodewyk Crowther (the son of the man who had challenged James Scott's monopoly of Vandemonian bodies) and solicitor Morton Allport established and maintained connections with British men such as Joseph Barnard Davis and William Flower. The latter, in turn, performed the 'higher task' of making knowledge about the Tasmanians and writing them into histories of humankind. In their relationships with Crowther and Allport, forged over a distance of 12,000 miles, Flower and Davis received the kind of important objects that boosted their own standing in scientific circles, enabling them to publish on the Tasmanians. In return, their colonial collectors gained what they valued most, which was acknowledgement that they were something more than forgotten men living in an outlandish place.

Neither Crowther nor Allport was particularly interested in Tasmanian skeletal material for the sake of science. Rather, each seemed to fit Fleming's description of 'native colonials'

> who had no choice but to return home [to the colony] after their studies
> abroad because nobody tried to detain them … [In compensation] they
> called on the old world to redress the psychological balance of the new,
> kept their affiliations with Britain in repair, and recovered the sense
> of belonging to a larger community in which colonial values did not
> prevail.[29]

Their sense of loss is sometimes poignantly obvious in these men's letters. William Crowther expressed it in 1869: 'I have always regretted having left London and the feeling gains … when I take up the periodicals and see what others have done in my absence'. And again in 1871: 'I have long been desirous to leave [the colony] and should have done so but like many others I have spent too much in the place and cannot … excepting at a great sacrifice get out of it'.[30]

The gifts he and Allport sent to Europe were weighed down with the significance of an expectation, sometimes crudely stated, that something would be received in return. That something was the relationship, no matter how tenuous, which offered these men prizes and awards and, in the Minutes of Proceedings in which their gifts were recorded, linked their names with those who were much more important contributors to science. There were also more nebulous rewards, such as the opportunity to drop an authoritative name into a local conversation and thereby boost their own status in the colony. In all these ways, local cultural authority was being forged.

William Crowther's Gifts

When William Lodewyk Crowther travelled from Van Diemen's Land to Europe in 1839 to complete his medical education, he took with him a large collection of Vandemonian fauna, dead and alive, which he sold upon his arrival in England for £300, an amount sufficient to allow him to complete his medical education in London, Paris and Brussels. Crowther greatly enjoyed his time in Europe, during which he also persuaded his young and well-connected cousin, Victoire, to elope with him. However, soon after their marriage, Crowther was called home to Van Diemen's Land, for his father had died and he was needed to take over his medical practice.

During the next forty years or so, Crowther was at the centre of the kinds of colonial activities that were available to a brash man who intended to make his way in this world. He maintained an extensive medical practice and engaged in entrepreneurial activities that saw him win fortunes—and lose them. He commissioned whaling ships to roam the southern seas, had prefabricated houses built and shipped to the Californian goldfields, obtained leases to islands in the Coral Sea from which the precious excrement of sea birds (guano) could be mined, and was elected to the colony's legislature, even serving for a brief and turbulent time as Premier (from December 1878 to October 1879).

This was the kind of life that should have been well documented, and some of it was: in the colony's press, parliamentary journals and other official records, and in articles Crowther contributed to the *Lancet*. But when he died in 1885, Victoire refused to accept what he had left her in his will, destroyed his personal and business papers, and packed up her two unmarried daughters and returned with them to England. Though Crowther's grandson, another medical William, wrote about his grandfather's life in a series of articles published in the *Medical Journal of Australia* during the 1930s, 1940s and 1950s, it is very difficult to find in Tasmania anything written in William Lodewyk Crowther's own hand about matters that were dear to his heart.

William Lodewyk Crowther, date unknown, by J. W. Beattie.
Courtesy of the Allport Library and Museum of Fine Arts,
State Library of Tasmania.

The letters Crowther wrote to William Flower, the Conservator of the Hunterian Museum at the Royal College of Surgeons, London, are contained in that
museum's incoming letters books. The first was written on 22 December 1863 in
response to an (unpreserved) letter Crowther had received from Flower, who had
heard of the colonial man's interest in whales and wrote seeking the skull of a
sperm whale for the museum. Crowther responded enthusiastically, writing that it
would give him 'much pleasure to make good the deficiency'. He introduced himself to Flower by saying he had

> been a resident in this Colony for the last 22 years actively engaged in
> the practice of my profession as a Surgeon and attached to the General
> Hospital and altho [*sic*] removed by great distance from your splendid
> Museum feel great interest in every thing [*sic*] tending to the advance-
> ment of Science.[31]

In what would become a pattern in Crowther's correspondence with Flower, he then moved on to talk of one of his sons, William, who would soon arrive in London to further his own medical education.

Young William had spent three years as his father's apprentice in the colony, two of them at its General Hospital, which was, Crowther informed Flower, 'a School not inferior to many larger ones in Great Britain'. Actually, it was nothing of the kind, except in the extent to which it offered its students opportunities to dissect. Crowther hoped that Flower would meet all of his sons over the following years, for there was more than one to educate in London, and that through them he would find that 'at the antipodes we have not degenerated physically or mentally, and still retain our affection for and attachment to the Institutions of the Mother Country'.[32]

This comment illustrates the resonance of metropolitan debates about human progress and degeneration for men who lived in what had, until mid-century, been a penal colony also populated by an Aboriginal race believed to be the most primitive on earth. European theories of human degeneration suggested that civilised people who lived in lands occupied by savages would degenerate, both physically and morally. Van Diemen's Land posed a special problem for men who believed that even the most civilised of human beings were capable of degradation.

Crowther also wrote that any attention Flower showed his son while he was in London would be reciprocated 'should any person reach Tasmania with a note from you'. It was the kind of *quid pro quo* that would become a feature of Crowther's correspondence with Flower over the following years. Yet there is also something endearing in Crowther's concern for his sons, which encompassed much more than a hope that they would become ornaments of their profession.

Over the next decade, Crowther made many donations to the Hunterian Museum, including the skeleton of a sperm whale measuring nearly 51 feet in length, the shipping of which would have been quite a feat. Crowther's name is listed in the catalogue of the museum's holdings of cetacea. His skeletal donations included the skull of 'a *Calf* sperm whale killed in company with the Mother'.[33] In return, Flower sent Crowther a set of the College's publications and, over the years, copies of his own articles on whales, also repaying his debt to the colonial man by making his three sons' lengthy stays in London both agreeable and profitable.

Almost all of Crowther's donations to the Hunterian Museum were skeletons of the huge fish and mammals from the southern seas that so fascinated European naturalists. In England, these bones were objects of analysis and discussion between eminent scientists, and Flower went on to make them into a personal specialty, publishing a number of scientific papers on the subject and becoming a world authority in the process. In large part, this was due to Crowther's continuing donations and the kind of detailed observations he wrapped around them. His words made them meaningful gifts. He was surprised to learn from Flower that 'so little was known' in scientific circles about the cetacea, and he delineated the differences between the species of whales, the 'Thresher', 'Black Fish' and 'Killer', informing him of their habits, preferred locations, the currents they followed, the size to which they grew, their different diets, and so on. He had obtained this information from the captains of his whaling vessels.

In return, Flower linked Crowther into metropolitan scientific discussions, sharing with him his own thoughts about pronouncements by the British Museum's Dr Gray on the skull of the blackfish Crowther had forwarded (Gray thought it was a new species, but Flower was doubtful). He told Crowther that, as an antipodean collector, he was in a position to assist in such disputes among naturalists, and asked for further opinions on the 'killer' that Crowther had forwarded. Was it part of the same species as that which inhabited the northern seas? He also requested more bones, while thanking him for his promise of the skull of a sperm whale and expressing his understanding of the problems involved in cleaning such a thing on board a whaling ship.

Finally, tentatively—just dropping the words between two sentences that deal with whales—he made a large and difficult request:

> I suppose it is useless to think of an entire skeleton (even of a female, or young one) as it could be only obtained from an animal cast ashore— & then at considerable expense—and I could not guarantee any large sum being given by the Council for such a specimen ... I suppose there is no further chance of obtaining a skeleton of a male & female [*sic*] one of the aboriginal ^{human} [*sic*] inhabitants, or a pair, male & female? ... If

Sir William Henry Flower, date unknown, by Elliott and Fry.
Courtesy of the Wellcome Library, London.

it falls in your way ever to get us an entire [whale] skeleton we should be most happy to have it, and perhaps I need not tell you as ~~a medical man~~ an anatomist, the importance of such skeletons being as perfect as possible …[34]

Crowther did ship off the entire skeleton of a whale, and was rewarded with one of the College's prestigious Gold Medals, which had the effect of linking his name with such eminent previous recipients as Thomas Huxley. He had considerably more difficulty in filling that request for 'a pair' of Aboriginal skeletons.[35] He informed Flower that he would not forget his 'hint', although '[o]ur Aborigines are *now reduced to about 5*'. Crowther said he would try to

*Studio photograph of Truganini, William Lanne and Bessie
Clark, August 1866, by J. W. Beattie after C. A. Woolley. The
other two people to whom Crowther referred as still living in
his letter dated 23 May 1864 were Patty (d. 1867) and
Wapperty (d. 1867). After their deaths, four at least of these
five people's skeletons came to lie in one museum or another.
Wapperty's fate is uncertain. Courtesy of the
Tasmanian Museum and Art Gallery.*

get an order sent to the Establishment where they reside (25 miles from here) that in the event of *serious illness* they should be forwarded to the Hospital where if they depart this life attention shall be paid to the prosecution of their osteological remains. I will make enquiries for I feel assured with no very great trouble a couple could be exhumed from their burial ground.[36]

However, the letter that contains these words, written in 1864, is revealing of where Crowther's main scientific interest lay, for it comprises six closely written pages, in which Aboriginal bones occupy a mere two-thirds of one page. The rest concerns whales. Three years later, he informed Flower that he would investigate the graveyard at Oyster Cove, though 'some little tact will be required as these poor people … have great repugnance at any disturbance of the remains of their friends'.[37] There is no further talk of human skeletal material in this correspondence until 1869.

Morton Allport's Gifts

Tasmanian lawyer Morton Allport was the scion of one of Hobart Town's society families. By the time of his death at the age of forty-eight, he had accumulated a large number of scientific honours to become

Morton Allport FLS, FZS, FRCI, Corresponding Member of the Anthropological Institute of Great Britain, Life Member of the Entomological and Malacological Societies and Foreign Member of the Royal Linnean and Royal Botanic Societies of Belgium, F.L.S., NSW, Vice-President of RST.

According to an obituary in Hobart Town's *Mercury* in 1878, these honours had the effect of making Allport the greatest naturalist in Tasmania. His main love was botany and, as a member of the Tasmanian Acclimatization Society, he tried to breed salmon from imported ova. Although Allport insisted these attempts were successful, visiting English novelist Anthony Trollope was less sure. When he

*Morton Allport, date unknown, by Charles Woolley. Courtesy of
the Allport Library and Museum of Fine Arts,
State Library of Tasmania.*

visited the Salmon Ponds in 1873, he said that Allport's 'enthusiasm [was so]
catching ... I found myself ready to swear, after hearing him, that there must be
salmon'.[38]

Like Crowther and others, Allport also assisted in gathering examples of
Tasmanian products to display at intercolonial and international exhibitions,
including the Exhibition in London in 1862. And most importantly, he was a
Fellow of the Tasmanian Royal Society, in which he served as Vice-President
(1870–78) and Curator of its museum (1870–75). In this position he learned about
museum exchanges, and what could be obtained in return for shipping Tasmanian
oddities to Europe.

Until 1869 there is little evidence that Allport was particularly interested in the Tasmanians, alive or dead. His thinking is illustrated in a diary entry he made in October 1853 while visiting Europe, as well-to-do colonial men did.

> At Armetiq fell in with original of Dickens' Mrs Jelliby, who wanted to know if I did not feel very like a returned convict, coming from a country which the inhabitants had stolen from the poor natives & then put them into confinement & used them in the most horrid barbarous manner instead of converting them (here was a Sally).
>
> I told the female that I was quite unable to argue the matter with her because she appeared to be well acquainted with facts which we poor ignoramuses in the colony were quite in the dark about, specially with reference to the barbarous manner in which the aborigines were treated, for that in the opinion of all people out there they were better treated & better off in every respect than they were in the helpless savage condition in which they were first discovered. I told her that I was afraid (with Dickens) that there were too many of the poor classes in England who were very far worse off than the aborigines abroad, but it did not suit the inclinations of would-be philanthropic ladies & canting ministers at home to exert themselves for their benefit as there would be some necessity for showing the proper feeling by more substantial means than talking. (After the unmerited insult she gave me I did not feel inclined to spare her.)[39]

This entry was the longest Allport made in his European diary. Being called to task as a representative of a convict society, for brutality towards the natives, had infuriated him.

Eighteen years later, Allport began to collect Tasmanian skeletal remains to ship to Europe, and for him, nothing but whole skeletons would do. He soon became the most successful collector in the colony, in terms of both quantity and quality. Whether his colleagues at the Tasmanian Royal Society knew the full extent of these activities is unknown. At this time they were arguing that such scarce material properly lay in their own national museum rather than in Europe, and Morton Allport's voice was loud among them. It makes his hypocrisy breathtaking.

In December 1871 Allport made his first donation. He shipped 'a case containing the Skeleton of a Tasmanian Aborigine' to the Anthropological Institute of Great Britain and Ireland, of which Joseph Barnard Davis was a member. This skeleton, he informed the institute in a letter dated 30 December, was 'perfect with the exception of the coccyx and two or three small bones of the hands'. He said he had obtained it from Flinders Island with considerable difficulty. The institute was extremely pleased to receive the first Tasmanian skeleton to arrive in Europe. Allport's gift resulted in his name being recorded in the Council Minute Book, among those of such men as Sir John Lubbock, Colonel Lane Fox, Davis and Flower, as well as other colonial collectors. Two years later Allport also sent to the institute two Aboriginal death masks, two casts of skulls, six stone implements ('the rudest made'), and some photographs of the natives.[40]

Allport's next two skeletons were sent to William Flower at the Hunterian Museum. One, he wrote, was known as Bessy Clarke, who had been between forty-five and fifty when she died. Allport had obtained her skeleton from its burial place at Oyster Cove. He regretted that some of her bones were missing—three ribs, several small sections of her wrists and one foot, and also the coccyx. The digging had been done carelessly. Perhaps it needed to be, for the settlement at Oyster Cove had not yet been abandoned, and the diggers may have been observed.

The other skeleton in this shipment, labelled 'Male No. 1', had been disinterred from Flinders Island. Obtaining this pair of skeletons, Allport advised Flower, had caused him 'no small difficulty'. He always emphasised the trouble taken in making his donations in this way, although disinterring these bones would have been far easier for Morton Allport than for others. He had connections with John Dandridge, the Supervisor of the Tasmanians' settlement at Oyster Cove, and also with the surgeon Joseph Milligan, who had been the Tasmanians' medical attendant on Flinders Island. These men knew where bodies had been buried.

Perhaps Morton Allport believed that talking of trouble taken would increase Flower's gratitude for these valuable gifts. In any event, he quickly moved on to speak of what lay, for him, at the heart of this particular gift, offering Flower yet another 'perfect' skeleton if, in exchange, he would return to the colony a particular Tasmanian skull that was rumoured to lie in the Hunterian Museum. Flower's reply was disappointing. He denied that the College possessed the skull, and all

Allport received in return for these two skeletons was a copy of the museum's catalogue, which arrived with a letter from the College Secretary, Edward Trimmer, rather than from Flower himself.

Allport abandoned the College, but not his collecting. He made his next gift to Joseph Barnard Davis, writing on 23 January 1873 to prepare England's premier skull collector for the surprise. It was, he said, a 'treasure for you in the shape of an Adult Male Skeleton of Tasmanian native all but absolutely perfect' and with 'every tooth in position'. Davis would have appreciated that full set of teeth. This skeleton was also from the graveyard at Flinders Island. A month later Allport wrote that he could scarcely make up his mind to part with it at the last moment, but then his better feelings predominated. This time there was no missing coccyx, and of the 106 bones in the hands and feet, which were always fragile, only some two or three were absent. It was, he boasted, 'altogether a most noble specimen', which he knew its recipient would value highly.

He was obviously excited to be corresponding with England's premier skull collector. The correspondence itself was almost sufficient reward for the gift, and he requested little in return:

> Please accept this as a present and expend anything you would have been willing to give for it in the articulating and figuring it our only bargain being that I am to have three copies of any publications in reference to it one for myself, one for our Royal Society's library and one for our public library.[41]

Writing some three weeks later, he also hinted that the local Royal Society would greatly value '[a]ny memorandum from you however short about the skeleton or about any peculiarity in any part of it' (a publication which would also include, no doubt, the name of the man who had made the donation). Then he wistfully enquired, 'Shouldn't I dearly like to see the result of your article and labours?' He said he wished to travel to England to experience the triumph first hand, but it was not easy to leave a busy solicitor's business and his 'good parents'.[42]

Allport's letters reveal that he feared the English collector would forget his antipodean correspondent, once the primary object of his link to Davis had been satisfied. He offered an enticement for the relationship to continue, telling Davis that he was working to collect every record of the lost Tasmanian race and might from time to time learn something worth communicating. That a vice-president of the Royal Society and the curator of its museum was only beginning to gather such material—when it was believed that only one Tasmanian remained alive—says much about the society's priorities until then. While men in Europe had been noticing and commenting upon this people's imminent extinction for decades, as had the local press, the society's Fellows had their minds elsewhere.

In return for this perfect skeleton, Davis sent Allport a copy of his *Crania Britannica*, together with a long list of questions about the provenance of the skeleton he had received. This insatiable collector either misunderstood the terms under which he had received this gift from Allport, or ignored them. He simply wanted more. Allport, who had neglected to provide this identifying information when he made the donation, was not the kind of ideal collector Davis valued. In the past, he had been inept in preparing animal skeletal material for the long journey to England.

Ever active in the interests of his own collection, Davis also probed to learn more about the skeletal material held in the Royal Society's museum. He must have found Allport's replies disappointing: the latter could not give a name to the skeleton he had sent. All he knew was that it had been buried on Flinders Island at the time that the natives were stationed there, but no record of the person had been kept.

As for the society's own collection, he informed Davis in a letter dated 8 August 1873 that it contained the skeletons of just three individuals, all 'known': Patty, a recent skeleton, dissected but never buried; Malabackaniewa, a nearly perfect male from Flinders Island, who had been 'the leading chief of the tribe roaming over the Southern end of the Island, where Hobart Town now stands'; and the 'greater part' of the 'last male' Aborigine, William Lanney. Of skulls, he wrote, the museum contained 'besides those of Caroline and Augustus nine specimens of undoubted authenticity', and seven of which he was not so sure.

The last of Allport's skeletal gifts was made to the Institut Royal des Sciences Naturelles de Belgique in Brussels in the same year, 1873. He had already sent the Institut many specimens of Tasmanian fauna, shipping off in the previous year a large number of skins and skeletons of mammals, including of thylecenes and wombats, as well as a number of birds, shells, lizards, ten bottles of insects and some material relating to the Tasmanians (a cast of a face and a skull, plus four stone implements). The Aboriginal skeleton he sent to Brussels was also in perfect condition. He informed the Consul-General of Belgium in Melbourne on 9 June 1873 that it had been 'obtained at considerable expense both of time & money … which will I am sure be appreciated as there are but three others in Europe and they were sent by me'.

THE TASMANIANS IN EUROPE: READING THE BONES

By the time Davis and Flower sat down to write the Tasmanians into their versions of the history of humankind, their subject matter was understood to be virtually extinct. The 'last' Tasmanian Aboriginal woman, Truganini, died in 1876, an event that was viewed as closing 'a remarkable epoch in Tasmania's history'.[43] This made the skeletal remains that resided in European collections very valuable indeed, and moved the men who had received Allport's gifts to the pinnacle of interpretive power where the Tasmanians were concerned.

Their texts entered a scientific culture in which earlier debates between monogenists and polygenists had changed. Although Darwin had not written of human beings in his *Origin of Species* (1859), he had demonstrated how species could change through a gradual and natural process of adaptation to their environment, and the implications for humans were readily seen. In 1871 he extended his argument in *The Descent of Man and Selection in Relation to Sex*. Turning an earlier form of monogenism on its head, he wrote that there had not been one single act of perfect creation from which all men had subsequently degenerated (albeit some more than others). Rather, 'all civilised nations were once barbarous' and humankind had 'risen by slow and interrupted steps, from a lowly condition to the highest standard as yet attained by him in knowledge, morals, and religion'.[44]

Shockingly, this lowly originating condition had not been human. Rather, humankind had descended from an ape stock whom they resembled in their physical structure, and with whom they shared some psychological characteristics, such as instincts, emotions and sociability. Darwin understood how horrifying his interpretation of human history would be to many, as it was to himself.

With this radical contribution to the debate on human origins and diversity, it might be thought that men collecting bones would concentrate on searching for skeletal evidence to prove or disprove Darwin's theory: seeking fossilised remains that exhibited incremental change from apes to human beings (as evolutionists would expect) or very ancient but always *human* beings (as degenerationists believed). Neanderthal remains were the only suggestive fossils that had been found at the time.

However, many of the arguments between monogenists and polygenists remained focused instead on marking the differences between contemporary races of humankind. Much of the debate was between men who subscribed to different kinds of monogenism: evolutionists ('progressivists') and degenerationists, and the Tasmanians served as a linchpin in their discussions. Progressivist Edward Tylor thought it simply unimaginable that the Tasmanians ('man at perhaps the lowest intellectual and industrial level found among tribes leading an independent existence, on their own land and after their own manner') could be the 'successors of higher ancestral forms', as the degenerationists argued. Rather, they were evidence of the progressivists' view, marking human existence at its lowest limit, from which point differences of development, rather than origin, could be traced.[45]

The Tasmanians also served as exemplars of what happened to such lowly people during the passive workings of natural selection, for they had failed to survive when the environment in which they lived changed.

Some men, however (among them Joseph Barnard Davis), stuck to the earlier polygenist view—and in dying, the Tasmanians provided 'evidence' to buttress this argument, too.

Dr Davis's Tasmanians: 'Entirely Obsolete'

When Davis sat down in 1874 to script his *On the Osteology and Peculiarities of the Tasmanian, a Race of Man Recently Become Extinct*, he had before him all the material an avid collector had been able to gather. This consisted of a notebook in which he had jotted down the matters he considered important when reading other men's direct observations about this 'curious and uncivilized people'; a number of papers that others had written on the matter; concrete examples of the Tasmanians' material culture; and, most importantly of all for him, his personal collection of twelve skulls, and the skeleton that he owed 'to the great zeal and generosity of Mr Morton Allport'.

Davis believed that in settling down to write the Tasmanians into anthropology, he was performing a kind of posthumous service for them, placing a 'lost' people in a position of permanence in humankind. (Truganini was still alive at the time Davis wrote, but she counted for little since she now could produce no offspring.) The Tasmanians would owe their place in history, Davis said, to the

> [t]ransactions of one of the most celebrated of the Learned Societies of Europe, which is always ready to embrace in its proceedings any valid contribution appertaining to the chiefest of human sciences—anthropology.

Davis wished to clearly establish the differences between the Tasmanians and the Australians, with whom he believed they had been erroneously confounded. This, he thought, would demolish monogenist arguments and support the view that these were two separate and distinct races of man. It is worth noting that not all polygenists placed the Tasmanians as separately as did Davis. Samuel Morton, for one, thought they belonged in the Oceanic-Negro family.

Though Davis referred to the cultural material that also lay in his collection, in *Osteology and Peculiarities* he gave precedence to the evidence provided by the osteological remains. He noted the rarity of Tasmanian crania, proudly putting on record the fact that he possessed the largest number in his own collection, although they amounted to only a dozen. Poor Blumenbach, he happily lamented, had never been so fortunate as to obtain even one.

The bones told Davis many things, some of which he had difficulty in expressing, leaving the impression that his readers must simply take him at his word:

> There is a peculiarity in the physiognomy of Tasmanians, which is also expressed in their crania, that has long impressed my own eye, but I did [*sic*] not know, whether I shall be able to describe it in words, so as to make it understood by others ... [it is] a particular roundness ... which manifests itself in all the features.

Measurements, it seems, were inadequate to the task. However, by measuring the skeleton he had received from Morton Allport, and comparing it against that of a young Australian man he also possessed, he found the kind of differences he sought. The bones of the Australian were gracile, while Tasmanian bones were robust, like those of Europeans. The Tasmanian was shorter in stature, his skull very prognathous, and the chin decidedly more prominent than those of the Australian specimen. In fact, Davis's skeletons showed him that the only point of resemblance between the two races related to the teeth, which were uniformly massive.

Most importantly for him, measuring the internal capacity of the respective skulls—which, he believed, was the most indisputable evidence of the essential difference between the two races—revealed that the Tasmanian brain was larger than that of the Australian. This, he felt, illustrated the probable intellectual and moral superiority of the Tasmanian. And yet the Tasmanian was extinct, while the Australian lingered on.

This was a problem that brought Davis into a discussion of the role played by environmental factors in explaining human differences. In his way of thinking, the fact that the Tasmanians possessed a larger brain than the Australians should have resulted in the Tasmanians having the more inventive culture of the two races. Yet local observations and artefacts indicated the clear superiority of Australian culture. For example, on the mainland, Davis noted,

> some remote Australian, more acute than his neighbours, observed that by throwing a light and crooked piece of stick, or wood from the hand into the air, it might be able to revolve in definite courses and strike certain objects according to the desire and aim of the operator

and so, in remote antiquity, 'those beautiful and wonderful Australian implements the bomerang [sic] and the [woomera] or throwing stick' had been invented.

This, Davis thought, confounded the opinion of those who 'base their views upon the migrations of human races, and their almost universal spontaneous diffusion'. It provided strong evidence to buttress his polygenist view that the races were different and immutable, for the boomerang existed *only* in Australia. No other hunting people (even the Tasmanians, whose island was so geographically

TABLE OF MEASUREMENTS OF ARTICULATED TASMANIANS,
BY
J. BARNARD DAVIS.

(ALL MEASUREMENTS IN MILLEMETRES).	No, 1761 ♂ aet. c. 30.	Anthrop Inst * ♂ aet. c 30.	R. C. Surgeons, England.	
			♀ aet. c, 25.	♀ aet. c. 25.
	mm.	mm.	mm.	mm.
1. Height of the Skeleton, from the Vertex to the prominence at the base of the Os Calcis	1640	1584	1612	1408
2. Length of the Vertebral Column, from the upper surface of the Atlas to the lower surface of the last Lumbar Vertebra	523	533	477	459
3. Length of the Os Sacrum, in a right line	107	107	89	95
4. Breadth of the Os Sacrum	99	92	102	99
5. Height of the entire Pelvis, from a line on the level of the top of the Cristæ Ilii to another on a level with the lower surface of the Tuberosities of the Ischia	175	192	185	151
6. Distance between the Cristæ Ilii, inside	234	234	243	237
7. Distance between the Anterior Superior Spines of the Ilia, inside	208	208	233	214
8. Transverse diameter of the superior opening of the Pelvis	107	109	114	120
9. Conjugate diameter of the superior opening of the Pelvis	101	104	104	99
10. Pelvic Index, or ratio of conjugate to transverse diameter, taken as unity	·94	·95	·91	·83
11. Transverse diameter of the outlet of the Pelvis, inside the Tuberosities of the Ischia	76	78	82	105
12. Conjugate diameter of the outlet, from the lower edge of the Symphises Pubis to the tip of the Sacrum	109	117	112	117
13. Breadth of the shoulders from the outside of one Acromion to that of the other	302	302	368	315
14. Length of the Humerus, extreme length	312	302	312	266
15. Length of the Ulna, extreme length	274	265	284	266
16. Length of the Radius, extreme length	251	246	265	234
17. Length of the Hand, from the upper arch of the Os Lunare to the point of the middle finger	167	178	190	208?
18. Length of the whole upper extremity	725	710	755	622
19. Length of the Femur, extreme length	463	434	458	388
20. Length of Tibia, extreme length	383	380	395	309
21. Length of Fibula, extreme length	370	360	338	317
22. Length of Foot, extreme length	215	234	231	177
23. Length of the whole lower extremity	893	875	898	743
24. Proportion of the length of the Arm to that of the Leg = 1·00, of No. 18-23	·81	·81	·84	·81
25. Proportion of the length of the Radius to that of the Humerus = 1·00	·80	·80	·84	·78
26. Proportion of the length of the Tibia to that of the Femur = 1·00	·82	·87	·86	·80
27. Proportion of the length of the Femur to the Stature	28	27·4	28·3	28
28. Angle formed by the arch of the Pubis	62°	68°	70c	92°

* Now in Nat. Hist. Mus., South Kensington.

Table of Measurements of Articulated Tasmanians, date unknown, by Joseph Barnard Davis (H. Ling Roth, The Aborigines of Tasmania, courtesy of Fullers Bookshop, Hobart).

close to that of the Australians) had discovered it. Europeans, he said, were incapable of even *using* it.

In addition, the Tasmanians lacked other cultural implements that the Australians possessed: they had no shields to protect their bodies in fights, no dogs to help them hunt, and no boats sufficiently seaworthy to carry them across Bass Strait. Davis solved what was, for him, a conundrum, by turning to phrenology and finding that the Tasmanian race must have had a defective 'inventive faculty'.

He also sought to demolish the degenerationist view of human history, as personified in the Tasmanian 'Conciliator', George Augustus Robinson, who believed the Tasmanians could become civilised. When Robinson had travelled back to England, he had taken with him a collection of skeletal material, which Davis subsequently purchased from his estate in 1867. Davis used this material to deny Robinson's view, and pointed in disapproval to the muddle that Robinson had made of his collecting. He had mixed up the Tasmanian and Australian material, and the journal he had written was in such a jumble, Davis charged, that Robinson's words could not be trusted. Witheringly, he dismissed the conciliator's observations about the Tasmanians, saying these were the opinions of a man who had

> no pretensions whatever, to be regarded as a man of science, or to be an accurate and exact observer and his writings are chiefly remarkable for that absence of definite and precise information concerning the curious people among whom he spent much of his life.[46]

Despite Davis's self-perception that he was a practical man merely seeking facts, some of his reasons for dismissing Robinson's views reveal that he had already made up his mind about the Tasmanians, theoretically speaking. In *Osteology and Peculiarities*, he charged Robinson with holding what he considered to be the futile conviction that the Tasmanians could be civilised and converted to Christianity. For Davis the fact of their extinction proved otherwise. It was an example of what became of primitive races when their environment was drastically changed, as it had been with the arrival of the Europeans, following which they had been forced to pass quickly through 'all the phases of human history', to resist and withstand ... contact with people the most different from themselves in existence, and, as an inevitable consequence, they have gradually dwindled away ... and ... are now become entirely obsolete. Absolutely this historical period has not been longer than one century.

Davis concluded his work by using the Tasmanians to undermine all monogenist arguments:

The races with which the systematists have united the Tasmanians in their arrangements of mankind, and the trivial grounds upon which these alliances have been based, form a subject of too insignificant a nature to deserve to detain us. Systems are generally built up by an incorporation of much that is imaginary; although they may at times be of some use in grasping the vast varieties of man. They are often unable to withstand an appeal to facts. All that can be said with truth is that the Tasmanians are not Australians, they are not Papuans, and they are not Polynesians ... the Tasmanians were one of the most isolated races of mankind which ever existed. That they were a peculiar and distinct race of people dwelling in their own island, and different from all others. And they have been one of the earliest races to perish totally by coming into contact with European people.

Dr Flower's Tasmanians: 'Degrees of Difference'

Three years after Davis published his *Osteology and Peculiarities*, and one year after Truganini's death, William Flower made his own contribution to European knowledge about the Tasmanians. It took the form of a lecture delivered in Manchester and subsequently published as *The Aborigines of Tasmania an Extinct Race*.

Like Davis, Flower drew upon information from local observers to provide a brief synopsis of the historical context of the Tasmanians' extinction. In Tasmania, he said, colonialism was attended by the 'usual difficulties' that arise when 'a country already inhabited by a different race from the new-comers' is colonised. In this case the difficulties had been exacerbated by the fact that a good proportion of the newcomers consisted of 'convicts of the most hardened and degraded type, who ... took to a roving and lawless life in the forests as bushrangers, or on the islands in the straits as sealers'.

Such apportioning of blame for the destruction of the Tasmanians would have been very familiar to his readers. Even a polygenist like Robert Knox was adamant that their passing had been a matter of brutality and murder, rather than an inevitable withering away.

Flower, like many colonists in Tasmania, sought to clearly distinguish between civilised and uncivilised Europeans. He wrote that through these lawless men, British colonisation had unleashed forces that illustrated 'the darker side of human nature'; and when the Tasmanians were removed from their island they had rapidly declined due to 'unfavourable climate, total change of mode of life, absence of all the interests and excitements of the chase or of war, and home-sickness'.

As well as acknowledging European culpability for the death of the Tasmanians, Flower shared with Davis a faith that these people's remains, in particular their bones, had a story to tell. He said it was now only possible to speak of the Tasmanians 'zoologically', to 'treat them as we do fossil animals, and rely chiefly on their bones for distinguishing characters'. In addition, like Davis, Flower felt a strong sense of responsibility to collect skeletal remains to ensure that the Tasmanians took their place in history. He hoped that

> the present occupiers of their land, who have profited so largely by their extinction, will spare no pains to search for and secure to science all that still remains of the race which they or their predecessors have been the means of destroying.

Where Flower differed from Davis was in his belief that the zoological history of the Tasmanians was a matter of continuing debate, rather than of settled fact. For Flower, the bones provided an opportunity to explore racial similarities as well as differences and so throw into question polygenist assertions of extreme and unchanging separation between the races of humankind. Flower asserted the evolutionist view, which was that the history of man had for untold ages 'presented a shifting, kaleidoscopic scene; new races gradually becoming differentiated out of the old elements [in a process of] constant destruction and reconstruction'.[47] He was uncomfortable with the very term *race*, which he thought extremely arbitrary; and he accused those who insisted that humankind comprised different species of being ignorant of the zoological meaning of the term.

Flower read Tasmanian bones with an eye to understanding what the bodies of a group of people who had been completely isolated from other races for thousands of years might illustrate about human evolution. It was this extreme isolation that made the Tasmanians important sources of knowledge to evolutionists. They were a people 'unaffected by all the complicated ethnological problems arising from a mingled influence of diverse and various races found among the nations of most other parts of the world'.[48]

This scientist argued that the Tasmanians had originally been part of one common stock, but that during the thousands of years in which they lived in a discrete ecological pocket isolated from the people on the islands around them, they had slowly transformed. For evolutionists, the lure of such isolated populations was the possibility that they provided evidence about which races were prior to, or derivative of, others. Were the Tasmanians a branch of the Melanesian race? Or were all Australians once like the people from whom the Tasmanians had derived, who had in turn been superseded by the present Australians? The evidence, he believed, would be found in a certain observable peculiarity of structure when the Tasmanians were compared to other races of human beings. However, these peculiarities were a matter of degree, of 'certain indications', rather than a matter of kind.

Flower doubted the very possibility of the project that occupied polygenists to construct a classification of mankind founded upon physical structure. Rather, he sought evidence in the bones for what it was that linked the Tasmanians to other people dwelling geographically close to their island.

This was a hard task, in the face of what appeared to be extreme differences, and it was important to understand what, exactly, was racially 'characteristic'. Flower worried about 'the present uncertainty as to the true classification of the varieties of human species ... the numerous cases that have to be dealt with of mixed or doubtful descent'.[49] In addition, he had discovered from his dissection of a young Bushwoman that bodily differences between races that were apparently at opposite ends of the scale where civilisation was concerned, were not so stark as others had portrayed them.

The difficulties were illustrated in an exchange between Flower and the Duke of Argyll, a biblical anthropologist who would have liked the Tasmanian bones to reveal signs of a common origin but a subsequent differential process of human degeneration from a moment of perfect creation. Argyll situated the Tasmanians at the centre of his argument when he said

> unless we are to suppose a separate Adam and Eve for Van Diemen's Land, its natives must originally have come from countries where both corn and cattle were to be had, [and] it follows that the low condition of these natives is much more likely to have been the result of degradation than of primeval barbarism.[50]

Argyll was actively engaged in debates with evolutionists. He travelled to Manchester to hear Flower deliver his lecture on the Tasmanians and afterwards wrote to seek clarification on Flower's view of their structural difference to other races.

> I think I heard you say that the skull of the Tasmanian is the only skull on which you could pronounce with certainty as to the race [to] which it belongs; that it is so separate from all others by marked characteristics that it can never be mistaken. On the other hand, I think I have heard you also say that there is no skull, however exceptional, which cannot be occasionally matched by individual skulls among other races.
>
> May I ask whether this last generalisation holds good of the Tasmanian skull as well as of others? That is to say, whether skulls with all the characters of the Tasmanian (square orbits, etc.) do not occasionally occur among the higher races?[51]

Flower responded by stating that the Tasmanians were indeed distinctive: no competent observer could mistake them, for example, for a European, because Europeans were a much mixed race. However, 'speaking zoologically', he stated that none of the varieties amounted to what he would call specific, a vew he contrasted with that of a well-known advocate of the polygenetic view of man's origin (possibly Joseph Barnard Davis), who had declared that the skull of a New Hebridean in his collection had such strong specific characteristics as could not possibly be found in an English graveyard.

For Flower, the characteristics of the Tasmanian skull revealed that the Tasmanians were most nearly allied to the Australians, Melanesians and the Negros, with whom they shared some characteristics, though others were singular. Humankind, he later wrote, was a matter of 'endless gradations of distinctive characters' and when anthropologists spoke of

> branches and sub-branches, varieties and sub-varieties, races, species &c ... these all are attempts to express degrees of difference connected by endless intermediate conditions, and passing insensibly from one to the other.[52]

Nevertheless, those few groups of humankind that had existed in conditions of isolation for lengthy periods of time bore a certain uniformity of character that applied to bodily, intellectual and moral qualities through the process of natural selection.

Evolutionists and other monogenists shared with polygenists a belief that civilised races were intellectually and morally superior to savage races. However, they differed in the degree to which they believed moral and intellectual qualities were linked to cranial capacity. While for Davis the link was direct and unchanging, Flower was generally careful to distinguish between the Tasmanians' 'capacities, intelligence and moral qualities' and their physical structure. He did not find the Tasmanians to be mentally or morally 'inferior to other uncivilised races' and said that he was disinclined to 'enter into the question of moral and intellectual character' of different races, for

> there is no subject upon which it is so difficult to obtain satisfactory evidence or to draw just conclusions. It is hard enough to do so with people about whom we have ample means of judging, but to attempt it with

savages, whose language is imperfectly understood, and whose ideas and notions are most difficult to appreciate, would lead us far beyond the scope of the subject I have undertaken ...[53]

Nevertheless, Flower found it hard to imagine human beings living in lower social conditions than the Tasmanians. He listed the defining aspects of their culture in a long series of negatives. The Tasmanians had no fixed habitation, no garments (though the climate in which they lived was far from tropical), no cultivation of the ground, no dogs. They had been 'quite ignorant of the potter's art', having no 'vessels for holding water, except pieces of bark or shells'. They had 'no kind of intoxicating beverage', and no tobacco. Their cooking was exceedingly primitive. Kindling a fresh flame was difficult for them. Their weapons comprised only a spear and a waddy, both of which were made of wood; and their domestic utensils consisted of stone axes and 'rude knives, baskets and fish nets', the Tasmanians being quite ignorant of the use of metal, or of bones for needles.[54] Their culture seemed to place them as the closest thing to man's lowest state, the point from which progress began. For evolutionists, this was a very fruitful point indeed.

European scientists' quest to plot human diversity was a congenial undertaking for many medical men, who brought to this work two special attributes. The first was a thorough knowledge of human anatomy, in both its philosophical and practical forms, which they had learned in a comparative way. The second was skill in dissecting human remains, including those that had been illicitly obtained. These attributes enabled them to contribute in a special way to debates about the peopling of the world. In particular, British medical men were well placed to collect the anatomical 'facts' that underpinned physical anthropology, and in Tasmania they collected rare skeletal material that was thought to provide important forms of evidence about the history of humankind.

The relationships formed through these desirable bones enable us to better understand the place of colonial collectors in these exchanges. Neither William

Crowther nor Morton Allport was primarily interested in scientific questions about the Tasmanians' place in the human scheme of things, but each had other interests that were well served through the links they made with metropolitan scientists. These relationships were complex: both scientific and personal, and driven by local as well as metropolitan imperatives. The quests to obtain Aboriginal bones and turn them into gifts were telling moments in the culture of medicine.

CHAPTER FIVE
DEATH AND DISSECTION, 1869

IT IS A FRIDAY EVENING IN MARCH 1869, AND TWO BODIES LIE in the General Hospital in Hobart Town. These men had grown old in this colony at the far end of Britain's world, and when they died as patients in the hospital a fortnight ago, no-one came to claim them for burial. This left them in the surgeons' hands. An offensive smell permeates the dissecting room. What is left of the men lies on two tables. For a fortnight now, by daylight and by candlelight, these bodies have been subjected to probing hands and instruments. They have been cut and torn, hammered and chiselled, snipped and sawn until they are no longer recognisable as individuals.

Not so the body carried into the dead house next door on this evening. The door between the rooms is open, and a new smell enters the dissecting room. Diarrhoea clings to this man's clothes. His arrival in the hospital is causing some excitement. People come to look and talk and touch. The words 'King Billy' hover in the air, where they are surrounded by laughter. Dan, the hospital barber, removes the stinking clothes and washes the body, then lifts it into a coffin. More visitors arrive, some with plans in mind. This man's blackness seems to fill the living with a possessive desire.

When all is quiet, one of the hospital's students, Bingham Crowther, arrives in the dissecting room. His father, who is a surgeon in the hospital, comes with him. By candlelight, these two go to work on one of the tattered bodies lying there. The older man makes a deft cut through the material that still holds Thomas Ross's skull to his body. He passes the skull to his son, and the two go into the dead house. Leaning over the body in the coffin, William Crowther makes a neat incision along the side of William Lanney's head. Then he carefully peels back the facial skin and cuts through the vertebrae beneath the skull. The surgeon lifts the skull out and Bingham hands him Thomas Ross's skull to insert in its place. Then the black skin is drawn over this substitute skull to assume something like the shape of a face once more. Thomas Ross could never have imagined playing such a part in a post-mortem minstrel show. Then the candle is blown out, and the living leave the room, taking William Lanney's skull with them.

Meanwhile in the dead house, Thomas Ross's skull lies uneasily beneath the foreign skin that now covers it. In an hour or so, the hospital's resident surgeon enters the room to check on William Lanney's body, which he has been instructed to guard. It appears odd. He lifts the head to take a closer look, and Thomas Ross asserts his presence. The skull shifts beneath George Stokell's hands, turning so that the bones of the face can clearly be felt through the back of the head. Panic ensues. The resident surgeon shouts for the hospital's steward and mounts a desperate search for Lanney's missing skull.

When it cannot be found, he contacts the Secretary of the local Royal Society, which he knows longs to possess Lanney's skeleton. The society's Fellows give Stokell a specific instruction. He is to cut off Lanney's hands and feet, to prevent Crowther returning to the hospital for the rest of the body, then put the lid on the coffin to keep these serial acts of theft a secret.

When Lanney's friends arrive to accompany his body to its funeral on the following day, they have heard rumours that he is no longer in one piece. They insist that the lid of the coffin be removed. Then the coffin is sealed and the public procession from hospital to St David's church begins. More than one hundred men join in, for this event marks the death of the 'last' Tasmanian man. The coffin is eventually lowered into the earth in the town's burying ground, while words such as 'salvation' and 'sanctity' and 'rest' are pronounced. Then the living—you could not really call them mourners—move away.

All is quiet until midnight, when the sound of a barrow being wheeled can be heard. Spades slice through earth under the impact of a serviceable pair of gardener's boots. The blades strike wood, the coffin is partially unearthed and its lid is chiselled off. Rough hands haul Lanney's shrouded body free. This is too much for Thomas Ross's skull. It drops to the earth and is kicked aside, while the hospital's gardener and messenger awkwardly thrust the body into a sack. Then the barrow is laboriously wheeled back to the hospital, which is uphill all the way.

Thomas Ross is about to get his revenge. His skull has landed on a nearby grave, where it waits until morning (the sexton's dog aside). A group of early visitors to the burying ground discover it there, and a wildfire of excited speculation is set off, which soon embroils the whole town. While George Stokell spends the day harvesting Lanney's bones on behalf of the Royal Society, Thomas Ross's skull sets about the business of destroying careers. It shakes a marriage, brings men and their science into ill-repute, and notches up a dozen other, less obvious little victories before the wildfire finally burns out.

This vignette may appear strange to readers familiar with those histories of racial relations in Tasmania that have made mention of the theft of William Lanney's skull. Thomas Ross has rarely entered them, except as a bit player, cast as 'the foreign skull', 'the unwanted skull', 'a neighbouring corpse', 'the skull of a European', 'a second cadaver of an indigent European', and 'the head of a dead white man'.[1] Historian Lyndall Ryan is the most informative. She gives the skull a name, Ross, and an occupation—he was a schoolmaster.[2] But just at this point, when there was a possibility for Mr Ross to finally enter history, he has instead remained an unexplored and ghostly presence in a hospital dissecting room.

Yet what happened to William Lanney was always an episode that intimately involved *two* bodies; and it was also always a matter of something other than race. What might we find if we return to the dissecting room and take each man as we find him, moving Lanney's body and story gently to one side so that others can also come into view? Refocusing in this way will place the daily practices in the Hobart General Hospital at centre stage, and allow us to attend to other contexts in which Aboriginal bones were collected for science.

IMAGINING MEDICINE IN COLONIAL TASMANIA

When William Crowther went to work in the hospital dissecting room that night, he was at the peak of his career as a surgeon, as well as a colonial entrepreneur who was in the process of running for election as the Hobart Town member of the Legislative Council, the colony's upper house of parliament. This was the kind of social position his father had sought to achieve following the family's arrival in Van Diemen's Land in 1825 and, as we have seen, it included agitating against Colonial Surgeon James Scott for changes to the colony's system of medicine, including who had access to the bodies of dead people. Over the following years, when writing to England, Tasmanian medical men boasted of their opportunities to dissect, which differed markedly from those available to British medical men. Edward Bedford, Scott's assistant who had been present at Mary McLauchlan's dissection, said each of the students at St Mary's, a private hospital he ran, had to attend not less than sixty lectures or anatomical demonstrations each year. When his students travelled to England to complete their medical education, they won prizes in anatomy.

In planning for the closure of the Convict Medical Department when the colony achieved responsible government in 1856, many of Tasmania's medical men understood their opportunities for practice would be greatly expanded, and sought to replicate, as closely as possible, the English way of doing things. It was one in which attachment to a charitable hospital, funded and governed by philanthropists, provided great benefits to those who were sufficiently fortunate to gain appointments to such institutions as 'honoraries'. This status gave them ready access to patients' bodies, alive and dead, while they maintained complete professional independence from both the institution and the government. In addition, they gained opportunities to provide their fee-paying pupils with clinical opportunities.

However, some Tasmanian medical men understood that such an independent system would be untenable in what had, until so recently, been a penal colony in which few philanthropically minded gentlemen existed to fund charitable institutions. Hospitals would necessarily continue to be financed by the government, though most men believed governments had no proper role to play in managing them.

Dr James Agnew had been one of the fortunate men with an appointment in the Convict Medical Department, in which he served as an Assistant Surgeon. In 1859 he jotted down his thoughts on how the hospital in Hobart Town should be run when the Convict Medical Department ceased to exist. His notes indicate a fine understanding of the kind of rivalries that arose between men holding different positions within hospitals. They enable us to understand some of the issues involved in the power play that was revealed in the furore surrounding the theft of Lanney's skull ten years later.

Agnew began his statement with a reminder that a hospital must be run in such a way that the interests of the sick were absolutely the first consideration. Professional interests were not to be compared with this. The fact that this statement prefaced his suggestions reveals that the opposite could be the case, and that he thought working out a system of authority in the colonial hospital would be no easy matter.

Like others, Agnew was uncomfortable with the thought that a government might manage a hospital. He therefore argued that the institution should be administered by a board comprising influential inhabitants and perhaps Hobart Town's mayor. In addition, rather than replicate the British system in which honorary medical officers ran the medical side of things, he thought the board should appoint one first-rate man as the hospital surgeon, who would have the general charge of the institution. To attract the right kind of candidate, and ensure that the odium attached by some members of the profession to the title of resident medical officer could be avoided, Agnew thought this should be a position that paid well. He suggested a salary of £500 per annum, which could be boosted if the hospital surgeon provided consultations to general practitioners (rather than to their patients directly, as this might alienate patients from their usual doctor and so cause professional concern), and if he earned fees from three pupils he taught in the institution. All up, this would give him a salary of £1000 per annum.

The reason Agnew gave for not appointing voluntary or honorary medical officers to this antipodean hospital was that in Tasmania such a scheme would attract few men; unlike in England, it would be no sure road to fame or fortune. He also had another concern. In a place like Tasmania, Agnew worried that a board comprising honorary medical men's fellow colonists would be unable to control the honoraries in the case of carelessness or dereliction of duty.

Teaching on the bodies of hospital patients was to play a major part in Agnew's scheme, as it did in England. However, at Hobart Town's hospital, he recommended that only the resident hospital surgeon should be allowed to instruct pupils in the institution, to ensure all students would work systematically at dissection. Such a school, he knew, would offer 'vast advantages' to its pupils, who would go home, that is to England, as accomplished anatomists, having had 'opportunities for Operative Surgery on the dead body that most English pupils would envy'.

In other words, this plan placed a salaried resident surgeon at the centre of things. Indeed, the only benefits private medical men would obtain were the opportunity to consult the hospital surgeon and a right of admittance to the hospital's 'Dead House'—which was, Agnew insisted, something 'we all should have'.[3]

When the hospital was transferred to the colonial government at the end of 1859, the system put in place was a hybrid of Agnew's suggestions and the English system. The members of the hospital's Board of Management were appointed by the government, rather than elected by subscribers. Where the medical side of things was concerned, one salaried but relatively junior resident surgeon was appointed, and in addition, the board elected a number of honorary medical officers, each of whom could teach his own pupils in the wards and hospital dissecting room.

This mishmash of systems caused anxieties from the beginning. Those private medical men who acted as honoraries gained what they had long been seeking, which was the opportunity to extend their practice in a multitude of ways within an institution whose medical side they virtually ran. However, the board that administered the institution was government-appointed, and its members had no medical expertise. This was something the honoraries constantly pointed out. They agitated for change from the beginning, and by 1869 had the opportunity to do so at board meetings, at which they were now represented.

Neither government nor hospital employees, the honoraries were a well-connected and vocal group. Some of them, like William Lodewyk Crowther, argued vociferously for the knot between government and hospital to be cut, and persisted in measuring the local institution against hospitals in England, where these men had trained. Most of all, men like Crowther wanted the hospital to increase in size, to a hundred beds, so that it would be recognised as a preparatory school by the Royal College of Surgeons. Then they would not need to meet the expense of sending their sons to England for so many years to complete their medical education. The best way of increasing the patient pool, Crowther argued, was to open the hospital's doors to the working poor, rather than just to those who were destitute.

A HOSPITAL DEATH AND DISSECTION

In this way, Thomas Ross entered the hospital, where he died in February 1869. Ross was seventy-four years old and in discomfort, but he would not have expected to die in that place—not from a recurring bladder infection, even in 1869. We do not know the circumstances in which he came to lie in William Crowther's ward in the General Hospital. Being a literate man, Ross may have read about Crowther's special skill where bladders were concerned and come to this particular surgeon in the hope of relief for the constant, nagging ache in his belly and the pain that made him afraid to urinate. He may have read many things about Crowther over the years—about his whaling ventures and guano islands, his shipments of timber-framed houses to California's fields of gold, and his spirited displays of rhetoric in a multitude of settings—but especially, he may have retained the memory that Crowther was a man who bent the rules. He admitted patients to the hospital who would normally be refused entry to that place.

Five years earlier, this had been revealed in a public stoush between Crowther and the hospital's resident surgeon at the time, George Turnley. Crowther had sent a woman to the hospital with a card of admittance even though, as her condition was not curable, the local Infirmary was officially the proper place for her. At a subsequent meeting of the hospital's Board of Management, its Chairman, Alfred Kennerley, argued querulously that if chronic cases were to be admitted, all the dying persons of the working classes would be sent in. Crowther's actions in this incident reveal that he knew how to get his way in the hospital by sidelining a resident surgeon, and that he thought little of doing so. He simply bypassed the resident surgeon and sent patients directly to the hospital's house steward, Charles Seager. This caused, the board heard from one of its members, Alderman Lewis, 'little feelings of animosity'. George Turnley interpreted it in tones more strident than that. He said that William Crowther treated him as if he were 'a mere nonentity'.

Crowther couched his defence in terms of common humanity, telling the Board that the woman spent her nights gasping for breath and that her husband was at his wit's end in trying to care for her and earn the money that kept the couple alive. To this, Turnley curtly responded, 'Let her be sent to the Infirmary'. Crowther angrily replied, 'She will die, and you … know it as well as I do'.[4] Three cheers for William Crowther, friend of the dying poor, we might think. Or we might wonder if he made such illicit admissions to the hospital for other purposes. He was a man with three sons to educate in medicine.

On 13 February 1869 Thomas Ross might have visited Crowther at his home, where patients with few funds were attended before the family's breakfast. This reception of the poor into a man's home for a free consultation seems strange to us, but it was a common practice at the time. After examining Ross, Crowther sent him to the hospital, probably by the indirect route he had successfully used before. Another slap in the face to a resident surgeon, this time George Stokell. Crowther's ward may have been particularly crowded, given his ongoing battle to increase the number of patients in the hospital. Indeed, he saw nothing wrong with having beds made up to accommodate them on the floor.[5] Thomas Ross was out of place in the hospital for another reason, too. It was a rare thing for a schoolmaster to be a patient there, among the paupers. Most patients were designated by such occupations as 'laborer', 'servant', 'seaman', 'groom'—or as the wife or child of such a man.

Ross lay in that ward hoping for a cure, for Crowther had made a name for himself in working on bladders. He was a mercifully speedy remover of stones. This was probably something he boasted about, for he was a boastful man. Four years later he would be writing to his London friend, William Flower, enclosing an article he wished Flower to forward for publication to the *Lancet*. It was a colonial contribution to a metropolitan debate about how best to remove stones from bladders, made by a man who felt a keen sense of the losses he had sustained in living so far from the centre of things. In the article, Crowther argued against a recommendation by Sir Henry Thompson, who was, he acknowledged, a great authority on the bladder, that stones could be crushed rather than cut out of aged and debilitated patients suffering from renal disease.

Crowther informed Flower that Thompson seemed to be 'totally unacquainted with the operation [methodomy] I have for years been performing in aged and debilitated people successfully'. He said he had worked on 'subjects with bladders in a high state of "chronic" irritation and inflammation', and these subjects, who had stones of considerable size, had ranged in age from sixty to seventy-five years.[6]

Crowther's article appeared in the *Lancet* eight months after Thompson's. Distance placed antipodean medical men at a severe disadvantage in debates in metropolitan journals. He argued for a combination of cutting and using the finger to dilate the prostate gland and remove a stone, on the grounds that this single, speedy operation would be less shocking to an aged patient's system than inflicting several episodes of crushing. Even if a stone were large, he insisted, his operation could be accomplished in two minutes, while lithotrity could take up to ten sittings. A week after this article appeared, a *Lancet* correspondent disagreed, writing that such a method was not supported by facts, as the prostate gland could not be sufficiently dilated, so the process involved laceration and the unnecessary loss of patients. Crowther, he suggested, had not kept up to date with surgical practice.

Was Thomas Ross one of the aged patients upon whom Crowther had performed his preferred operation? When the wardsmen carried his body into the dissecting room, Crowther was probably the first to open him for inspection. His surgical training would have encouraged him to investigate the results of his work, post mortem. After that, Thomas Ross became the students' thing, for no-one came to claim his body and so remove it from the surgeons. A fortnight passed before we hear anything further about him.

DEATH OF THE LAST MAN

On 3 March, when Thomas Ross's limbs had become the most useful, because the least decayed, parts of his body, a man variously called 'King Billy' or William Lanney died in an inn located in the seedier part of Hobart Town. Lanney was thought to be the 'last' Tasmanian Aboriginal man. He was also a member of Tasmania's peripatetic whaling community.

Lanney had been born near the Coal River, some six years after Mary McLauchlan had arrived there to work on the Nairnes' property. He was one of five children in a family that was, in 1842, the last to be persuaded by George Augustus Robinson to leave the mainland and move to Flinders Island. When Lanney was twelve, he was sent to the Queen's Orphan Asylum, and from there was apprenticed, in 1853, to a farmer at Clarence Plains. After that, he spent most of his time on whaling vessels, such as the *Aladdin* and *Runnymede*, or in Hobart Town's inns in the company of his fellow crewmen, and sometimes visited the settlement of elderly Tasmanian women at Oyster Cove, one of whom, Truganini, was said to be his wife. In 1864 Lanney argued for increased supplies of food for this community, stating, 'I am the last man of my race, and I must look after my people'.[7] However in the main, his time was spent whaling.

Contemporary writers said that Lanney was popular with his fellow whalers, being good-natured and jolly. Historian James Bonwick tried several times to meet him between his whaling voyages, in a quest to learn more of the Tasmanians, but was left lamenting that he could never find Lanney sober. This gave substance to his understanding that alcohol played a large role in the decline of the race, upon whom its blight had fallen.

Tasmania's whalers formed the crucial basis of an industry that had, as recently as the 1840s, been a lucrative one for the colony's entrepreneurs, among them William Crowther. By 1869 the industry was in decline, for other sources of oil were more easily obtained, though men like Captain Bayley, the owner of the *Runnymede*, upon which Lanney sailed, were still important whaling-ship owners.

Whaling was a dangerous and mythologised occupation. It forged a special kind of community out of people from very different backgrounds, at sea and also on shore between voyages as the whalers spent their earnings in the kind of boisterous behaviour that was decried by the respectable people of Hobart Town. Lanney died in his room at the Dog and Partridge Inn on 3 March 1869, during such an interlude between voyages.

In stark contrast to the non-event of Thomas Ross's passing, Lanney's death resulted in much public comment. The colony's press immediately printed potted histories to help people make sense of the rapid passing of a race. John Davies, the influential editor of the Hobart Town *Mercury*, expressed the view that such an event was 'not likely to stand unique for any very lengthened period in the history of British colonisation' and pondered the strange fact that 'wherever the white man has set his foot, in whatever quarter of the globe civilisation has been implanted, there the savage races have begun rapidly to degenerate', and to die out under the shadow of the 'pale-faces'. 'The death of Tasmania's "last man"', he knew, provided 'an abundant field for the speculations of moralists, philanthropists, and *savants*, which, doubtless, will be fully availed of both in England and the Colonies'.[8]

Others making history of Lanney's death believed that extinction had been a much more active process. They argued that the Tasmanians had been *made* to disappear, and thought it worthy of note that this state of affairs seemed more welcome to scientific men than others. William Crowther was among those who believed there was nothing inevitable about this people's fate. When writing later to William Flower, he said that Lanney's skeleton, 'the *skull in particular*', exhibited to 'a most striking degree the improvement that takes place in the lower race when subjected to the effects of education and civilization'. The '[m]ental development is well marked when the skull is compared with that of his ancestors, and the bonus is [that] the trunk and extremities will give evidence of much muscular power a circumstance quite at variance with the history of his race'.[9] Three months later he repeated the point, writing that Lanney's skull, when compared to 'its wild cousins', 'makes … a very valuable specimen', which could result in a 'very good text from any person inclined to take up the measurement of races as a popular subject'.[10] This opinion situated Crowther among those who believed 'progress' could be instilled in the Tasmanians.

Lanney's death set Tasmania's *savants*, several of whom were medical men, quickly into action, for it turned him into a rare collectible. The desire to possess the skeleton of the last man of an extinct race seemed irresistible. Such an object would guarantee a collection's continuing and unique importance, for extinction meant that the Tasmanians could now only be known through such physical remains. It placed them more firmly than ever as a people of the past, who could be mapped and interpreted as inanimate objects.

William Crowther made an official request for Lanney's skeleton on behalf of science. He believed he already had a firm agreement from the government that the next Tasmanian who died would be his to anatomise on behalf of London's Royal College of Surgeons, for two years earlier when a woman named Patty had died in the hospital, though he had been promised her body, it had instead been given to the Fellows of the Royal Society. Now, Crowther expected the government to honour its word and make Lanney his. But just to be sure, when he heard of the death, he quickly wrote to remind the Colonial Secretary, Sir Richard Dry, of that earlier promise. Simultaneously, he sent several police constables to remove the body from the inn where Lanney had died and bring it to the hospital, an institution in which he had some authority.

Lanney was placed in the hospital's dead house, a room reserved for laying out the dead to await families or friends to claim them for burial. The folding door that separated this room from the dissecting room was meant to make all the difference in the world about how a body was treated in the hospital, post mortem, a distinction that was a matter of shared understanding between medical men in the British world.

When Dry received Crowther's pointed reminder of his earlier promise, he was in a difficult position, for he had also received a speedy request for Lanney's body from the Royal Society, of which he was a member. In a note penned by Morton Allport, the society argued that the bones of the last man properly belonged in their 'national' collection. Dry knew, though, that Crowther was not the kind of man to whom a promise could easily be broken. Nor was the Royal College of Surgeons an institution to sneeze at. So he sought to negotiate his way out of this dilemma by asking the Royal Society's council to reconsider their request. Might they find it compatible with 'the interests of our local institution' to waive their claim in favour of the Royal College of Surgeons?

Dry might have thought the Fellows would give up their claim, for the society was dependent on an annual government grant of £5000. However, his request carried no weight with the men who met to discuss it. Three of them (James Agnew, the society's honorary secretary, Morton Allport and another Fellow) formed themselves into a sub-committee to draft a response in which they informed Dry that the society had the 'right' to possess the skeleton of the last man, for it would 'complete the series', as they already possessed a woman's skeleton. They argued that Lanney was an 'essential element of a national collection', and warned of repercussions should this particular skeleton be 'lost to the country and its museum'. During the latter half of the nineteenth century, colonial scientists

December 29, 1866]   TASMANIAN PUNCH.

THE WORKING MAN'S FRIEND
WORKING MAN:—Well Doctor, You have left us I see!
DR. C——R:—Yes my good Fellow—I think the lance of the whaler, and the lancet of the Surgeon pays me much better than politics.

The Working Man's Friend, 1866, by J. M.' (Craig, Mr Punch in Tasmania, courtesy of Blubber Head Press). In his surgical practice, Crowther was known as a friend of the poor. It was a designation he used to his advantage. This sketch was made at the time of Crowther's sudden departure from the House of Assembly three years earlier, angry at Colonial Secretary Richard Dry for not appointing him to a ministerial position. In this image, Crowther wears the top hat and coat that became a brief focus of attention during the enquiry into Lanney's mutilation. Some witnesses wondered whether Crowther could have smuggled Lanney's skull out of the hospital beneath these items of clothing.

were increasingly mounting arguments that reference or unique materials should be housed in local museums. (Although, as we have seen, Morton Allport was not always so patriotic where collecting was concerned.)

Agnew signed this letter to Dry in the everyday way, as the 'obedient servant' of the Colonial Secretary. In fact, he was nothing of the kind. These men were part of a tightly knit group who simultaneously occupied positions in government, the Royal Society, and on the General Hospital's board. Perhaps not surprisingly, given these circumstances, Dry accepted the society's assertion of local priorities. However, he was obviously suspicious about how William Crowther would react when he heard that his request for this body, too, was denied. So Dry sent a quick letter to the hospital's resident surgeon, George Stokell—the only medical man in the institution who was a government employee—to instruct him to protect Lanney's body until decisions about a post-mortem examination had been made. The Colonial Secretary made no mention of a burial, which was strange, given community expectations that the passing of the last man would be marked by a formal ceremony.

Crowther, however, was a force to be reckoned with, in the competition for Lanney's bones, as in everything else he did. When he heard about the contents of this letter, he understood that his request for the skeleton had been less persuasive than that put by the guardians of the local collection. Soon afterwards the serial work on William Lanney's body began.

SCANDALOUS BODIES

Two government enquiries followed the discovery that William Lanney had been resurrected from his grave. The first was held on the morning of Monday 8 March, following the weekend drama of the funeral, a deputation to Dry by Lanney's friends to insist the government investigate rumours his body had been 'tampered with' in the hospital, and the discovery of Ross's skull and Lanney's empty coffin. This was a brief ministerial enquiry, from which the press was excluded. Dry limited it to questions about the disappearance of Lanney's skull while his body lay in the hospital's dead house, and sought evidence from just four of the institution's employees (the resident surgeon, the steward, the barber and a messenger). Each pointed, as Dry knew they would, to Crowther and his son being the culprits.

On the basis of this enquiry, the government claimed to have found *prima facie* evidence of the Crowthers' guilt. Dry wrote to inform Crowther that he had instructed the hospital's board to bar him from the premises immediately. This swift instruction revealed two things: first, that the Colonial Secretary believed it was within his power to *instruct* a nominally independent hospital board; and second that Dry was aware that the hospital still contained evidence of Stokell's post-resurrection work on Lanney's body. Banning William Crowther from the premises, he hoped, would deter him from exposing this to an eager public.

By noon Alfred Kennerley had called a special meeting of the hospital board to implement the government's decision. The board added a further punishment of its own, banishing Crowther's son, Bingham, who was a pupil at the hospital. You can almost see both government and hospital board hopefully washing their hands of the affair. No chance now that Crowther or his son would find Lanney's bones drying out on the roof of the hospital.

But they had seriously underestimated William Crowther. He quickly pointed out, in public, that in this enquiry he had been tried and found guilty in absentia, which was not the kind of process an Englishman should expect. As for hushing up the fact that the hospital was a site of ongoing mutilation, he put paid to that by thrusting his way into the institution, hammering down the door, and revealing, in an advertisement published in the *Tasmanian Times*, what he found there. The room, Crowther reported, was 'a complete charnel-house; *the bones had been removed*, but masses of fat and blood were all over the floor and upon a large box which had been used as a table. In fact, the stairs, landing, and the room gave evidence of the hurried and indecent manner in which the remains had been disposed of'.[11] So much for keeping a lid on things.

Within three gossip-filled days, it was obvious a further investigation could not be avoided. Dry established an official Board of Enquiry, appointing William Tarleton, the Recorder of Titles, and two merchants, Charles Colvin and Isaac Wright, to conduct it. The wording of the announcement suggested to at least two of these men, Colvin and Wright, that they were empowered to enquire into *all* of the episodes of mutilation where Lanney's body was concerned: skull, hands, feet and the rest of the body. In fact, Dry intended questions to be asked only about the removal of the skull. There was simply too much at stake for government and hospital in an investigation that roamed more freely.

This second enquiry was open to the press, and during the three days that the board sat—and for some time afterwards—these events were replayed and reflected upon in press reports, editorials, advertisements from interested parties, poems and letters. While the board was empowered to call witnesses, Crowther refused to attend on the grounds that during the first enquiry, the government had already taken, and published evidence without him being present to defend himself. Worse still for the government, those who did appear and speak over the next few days readily wandered into territory that Dry had sought to place beyond the scope of the enquiry, and this revealed just how easily an individual person became a thing for the surgeons in Hobart Town's premier hospital.

OBTAINING SUBJECTS FOR DISSECTION

Reading the evidence presented at this enquiry, you can feel the increasing discomfort of the three men who posed questions to witnesses, as they began to wonder about the fact that there had been two other dead bodies lying in the hospital the night before William Lanney was buried. These men wished to understand precisely whose bodies these were, and how they had become so readily available for the institution's students to dissect. But it seemed to be impossible to obtain straight answers to their questions. It was not always that those who worked at the hospital and were familiar with its practices set out to deliberately deceive the enquiry—though some, like George Stokell, did—but that for them such work on bodies was mundane and they could not see the sense of the questions. The record of what was said reads like an exchange between cultures, in which questions were misunderstood, and the meanings of replies made no immediate sense to those who heard them, for things that appeared to the men on the Board of Enquiry as outrageous behaviour, seemed to those who worked at the hospital merely a matter-of-fact part of everyday practice.

Reading the transcript of what was said—which every literate person in Hobart Town could do, for it appeared seemingly verbatim in their daily newspapers—it quickly became apparent that it was nobody's job at the hospital to ensure that those who died as patients there went to their graves in one piece. Tasmanians of British descent began to wonder what such activities suggested about their social institutions, for cultural identity was always a prickly matter in this place.

In an effort to establish the provenance of the bodies in the dissecting room, Tarleton began by questioning George Stokell. He asked him directly, 'Whose bodies were they?' Stokell said he was not prepared to give the men's names, needing his book to refer to. Tarleton next asked Stokell whether the two white men had been removed to the dissecting room as subjects for the students to dissect, and to confirm that they were bodies of patients who had died in the institution and had not been claimed. The resident surgeon replied that they were. This was the first the Board of Enquiry had heard about what happened to the bodies of those without 'friends' who died in the hospital.

Tarleton, Colvin and Wright were becoming increasingly uneasy as they stumbled onto information they had not expected to find. They tried to elicit the exact identities of the bodies in the dissecting room, unable to believe that their erasure could be as complete as it seemed. They got nowhere with the resident surgeon, who insisted, with a kind of disdain, that he rarely went into the dissecting room. His role in the hospital was a far cry from the pivotal one that had been envisaged for a resident surgeon in James Agnew's plan. Nevertheless, Tarleton persisted. Finally understanding the significant difference between a dead house and a dissecting room, he asked whether any other corpses lay in either place and was informed that there were none in the dead house, but two lay in the dissecting room. The questions and answers that followed this admission reveal the cultural niceties of each man's position and how easily *some*body became an object in a dissecting room, at least to the resident surgeon.

Q. What bodies were they?

A. They were the bodies of two patients who had died in the hospital.

Q. Were they male or female?

A. Two males, Sir.

Q. These people who were in the dissecting room, of what persons were they the bodies?

A. Of two patients.

Q. Whose bodies were they?

A. I don't know, Sir; that is, I cannot say for certain. One belonged to Dr. Crowther, and the other, I think, to Dr. Bright.

Q. That is a somewhat curious expression, isn't it?

A. You asked me whose bodies they were. I suppose they were nobody's. I don't know their names.[12]

Enlightenment was sought from the next witness, the hospital's steward, Charles Seager. In a hospital, a steward played the part of an interstitial figure. In status, he was placed beneath the resident surgeon, and above everybody else who was directly employed by the institution. Seager was a literate man, a keeper of records, who had occupied this position for many years. All non-medical hospital employees deferred to him, and his approval or collusion was necessary when those above him in the hospital's hierarchy—the resident and the honoraries—wished to act in ways that stepped outside or stretched the rules, as was the case when William Crowther admitted unsuitable patients to the hospital.

Seager was very skilled at giving only the minimum amount of necessary information in his replies to the enquiry. He spoke about these events in a way that had the effect of removing him from the scene of the action, and seems to have had little respect for George Stokell. ('Dr. Stokell came to my quarters—about 9 o'clock. He said that Dr. Crowther had been in the dead-house, and that the head of the blackfellow was gone. He seemed rather excited about it'.)[13] Seager's answers to questions posed were confined to a series of affirmatives.

Q. Is it a usual thing for bodies that are not claimed by the relatives to be taken to the dissecting room?

A. Yes, Sir.

Q. For the use of the medical officers and students?

A. Yes, Sir …

Mr. Colvin: When you got the body of the white man did you not register his name?

A. Yes, Sir.

Q. You know his name?

A. Yes, Sir.[14]

At this chance to establish Ross's identity and write him into the records of their enquiry as something more than a spare skull, the board hesitated. Perhaps they remembered they were not meant to enquire into these matters; or perhaps they merely wished to be reassured that *someone* in the hospital was keeping a record of those who died there and of what subsequently became of their bodies. It was, after all, Tarleton's job as colonial Registrar to care about the recording of names.

The next hospital employee to be called to give evidence was 'Dan the Barber', also named by the reporter for the *Tasmanian Times* as 'John Seugrave (a most stupid Irishman whose brogue was so broad as scarcely to be intelligible)'. In his replies to questions posed, Seugrave was more voluble than the previous respondents had been, sometimes too much so for the board's comfort. William Crowther had claimed that he had only been at the hospital that night to assist his son (and pupil) to dissect an arm, and Tarleton asked Seugrave whether it was a usual thing for students to be dissecting at the institution after dark. The barber replied 'No, it is not'. But instead of leaving it there, as Seager would have done, he expanded. 'They can come any time they like, but they come mostly in the day time. They used to be always working there', he said. When asked about the identity of the bodies in the dissecting room, Seugrave responded:

A. I don't know. One, I think, belonged to Dr. Crowther.

Q. I don't want to know that. There seems to be an impression you people at the hospital have, about the ownership of bodies. I want to know what their names were?

A. I could not tell that, Sir …

Double-checking the answers he had been given by Stokell and Seager, Tarleton asked Seugrave:

Q. All the bodies of people who die, and are not claimed by their friends, are put in there to be dissected? Are they not?

A. Yes, Sir. When there are no friends to take them away.

The *Mercury*'s reporter recorded that the Chairman then remarked that the hospital must be 'a good school for surgery, as there were no doubt many such bodies'.[15]

Later, when asked precisely what the students were *doing* with the bodies in the dissecting room, the barber answered in a matter-of-fact way, saying 'I could not say what it was they were operating on, whether a leg or a body. I think it was a body, and it had its 2 arms on, and—and—and—I think it has them on still'. Others in the room laughed, but John Seugrave was trying his best to make the board understand how difficult their question was: 'There was an arm in the dissecting room, but [I] cannot say to whom it belonged. Cannot say whether the doctor was operating on the loose arm or upon the body'.[16]

This man had let the cat out of the bag, for of course, after being dissected in that room over two summer weeks, Thomas Ross was no longer recognisable as a man, nor was he in one piece. By the time he entered William Lanney's body and story, he had already been reduced to little more than a dismembered skeleton that bore the tattered remnants of other bodily material, which was something most men giving evidence at the enquiry sought to hide.

So far as Tarleton, Colvin and Wright were concerned, it was all becoming somewhat like Dr Frankenstein's laboratory. They wondered how it was that in a dissecting room a man could lose so much more than his limbs. Had they been medical men, they would have understood that successive erasure of identity from the moment of death. By the time a man wrote up his work, the process was complete. Human beings had been reduced to their gender, age, cause of death and anatomical peculiarities.

This official enquiry also became a kind of gladiatorial battle between the medical men who were its two main protagonists. Crowther's refusal to appear meant that he had no voice there, though he was very vocal in other forums. After reading Stokell's evidence at the first, ministerial enquiry, he had taken out an advertisement in the *Tasmanian Times* (heading it 'Et Tu! Brute!!'), in which he painted a horrifying picture for the public of Stokell's butchering activities following the robbing of Lanney's grave. Crowther wrote of following a bloody trail left by the wheelbarrow's tracks, which led him from the burial ground back to the hospital. Such an amount of blood, he informed readers, could only be accounted for by Lanney's body having been quartered. He also exposed Stokell's subsequent work on Lanney's body to public view, as we have seen.

In a coup de grâce, Crowther accentuated the professional distance between himself and the hospital's resident surgeon by claiming to feel ashamed of Stokell's behaviour. This made George Stokell stutter with rage. In a response, he wrote: '*He ashamed of my conduct!!* I am quite capable of bearing or repudiating any odium attached to the matter, after my explanation, and he will have no time to spare for being ashamed of any one's conduct if he occupies himself by blushing for his own'.[17]

Until the Friday on which all this body work had begun, the relationship between these two men seems to have been that of an established surgeon with a younger colleague. Crowther was one of the most successful Tasmanian medical men of his day, and Stokell was junior to him in every way. He was merely a government employee in the hospital, holding a position that carried few of the advantages James Agnew had envisaged for it. Stokell also owed Crowther social obligations, having travelled to England to complete his medical education in one of Crowther's vessels; moreover, he often took tea in the Crowther family home.

Not surprisingly, Stokell had been worried about the instructions issued to him by the Colonial Secretary to guard Lanney's body. The letter informed him that, pending further instructions as to post-mortem examination and the disposal of the remains, he was not to allow any mutilation of the body of the Aboriginal native, 'Billy'. Stokell understood the ambiguity encompassed by those words: no 'mutilation' but a possible 'post-mortem examination'. No doubt, two years earlier, he had been following such further instructions when dissecting Patty's body for the Royal Society, though it had been promised to Crowther for the Royal College of Surgeons.

Clearly worried about Crowther's reaction this time, Stokell talked to others who were also hospital employees, the steward and the dispenser, and then on the Friday afternoon he also consulted one of Crowther's fellow honorary medical officers, Dr Brooke, who advised him to watch Crowther carefully. Stokell decided to speak to Crowther directly, and ran to catch up with him as he passed by in the street. This informal conversation outside the hospital had been the turning point in the relationship between the two men. It was reported a week later when Stokell answered questions at the official enquiry, following the publication of Crowther's 'Et tu! Brute!!' advertisement, which had exposed Stokell's grave-robbing activities and butchering of Lanney's body. When Tarleton asked what Stokell had said to Crowther in the street, Stokell replied:

A. He asked me what would be done with the body of Lanney, and he said he wanted the body ... I told him he could not have it.

Q. Did you tell him why he could not have it?

A. Yes, I told him what the Colonial-Secretary's orders were ...

Q. What reply did he make to that?

A. He said the Colonial-Secretary had given him the body or promised it to him long ago ... he said he would have it. He said, when he is buried, Stokell, we can get him that way. He said, you look after him and see that he goes away all right, and we will get him then ... I told him he would have some difficulty in getting him out of the churchyard, as there were two dogs belonging to the sexton.

Q. Well! What did he say to that?

A. He did not say anything at all. I told him the Royal Society had the best right to it, and he said I was a fool to keep it in a paltry place like Tasmania, when it ought to be sent to a place like London ...

Q. What further took place?

A. Oh! He asked me to come to his house ... He said the hour I was to come to his house at was 6 o'clock. He said come to my house; come to tea ...

This conversation is very revealing. Though only Stokell's version is available, Crowther never denied it. These two medical men shared an understanding of the niceties involved in robbing graves and the need to be wary of sextons' dogs; neither of them believed Lanney's body would or should lie for long in its grave, and William Crowther initially assumed that Stokell would understand his own sense of entitlement to the body.

But having been denounced in that vicious advertisement, Stokell knew he had nothing left to lose at the enquiry. He repaid Crowther by denigrating the manner in which the senior man had removed Lanney's skull, saying that when he had returned to the hospital that night and found the foreign skull beneath Lanney's skin, he could easily see what had taken place, for Crowther had made a mess of the job. Adding insult to injury, he went on to suggest that such a thing could easily have been accomplished by the hospital's barber, though he had no knowledge of surgery at all, or even by a butcher.

The Board of Enquiry sat for only two days. When Colvin and Wright sought to extend its terms of reference beyond enquiring into the theft of Lanney's skull, Richard Dry refused their request, and the merchants resigned. The *Tasmanian Times* used language that harked back to the Burke and Hare murders, which had been the cause of Robert Knox's downfall, accusing Dry of having 'forbidden and burked inquiry'.[18]

Those two days of questions and answers had done more to fan speculation about these events than to end it. Crowther's many friends and political allies petitioned the colony's Governor and held large public meetings in his support. But he himself made a mistake in foolishly suggesting in the press that George Stokell, rather than he, may have stolen Lanney's skull—which gave Dry an additional opportunity to direct the focus of the debate. He had the Chief District Constable launch a police prosecution against Stokell for stealing the skull, at which Crowther had no choice but to appear and give evidence. Though the charge against Stokell was, of course, not proven, this prosecution was no good thing for the resident surgeon. To be accused of mutilating the bodies of the dead did nothing for a medical man's career prospects. But by now nobody except Stokell himself really cared about his career. It had become a kind of collateral damage in a much larger game, and he was left fighting desperately on his own behalf.

The police prosecution was pure theatre for everybody else—even Crowther, who managed to slide out from a few sticky moments by refusing to answer a great many questions on the grounds that he might 'criminate' himself. Of course, this defence revealed that he was the man who had stolen Lanney's skull. But by this time, everybody knew that, anyway.

BECOMING TASMANIAN

As news of the serial post-mortem attacks on Lanney's body began to circulate, history-making about the death of the last man receded from the press, to be replaced first by excited speculations about the mystery of the thing, and then by sometimes anguished reflections about what it meant to *be* Tasmanian—for that was now the nomenclature for the Europeans who lived in that place—in the face of such events.

Some thought it was more the manner in which Lanney's bones had been taken than the fact of their taking, which formed the core of the problem, for mutilating the dead violated a sense of decency. They preferred their science in the abstract, clean. 'Job Muggs', a prolific letter-writer to the press, was usually highly critical of British colonisation. However, he understood that science needed Aboriginal bones. He criticised the process by which they had been collected, rather than the fact of the collecting. Why had the scientists not disinterred bodies from the graves at Oyster Cove and Flinders Island, he asked, or gone about the business without 'mocking and travestying the most solemn Christian rites and ceremonies and outraging the laws of humanity and even those of decency'?[19] Waiting a decent interval, then resurrecting the dead from more distant burial grounds, it seemed, mocked nobody. John Davies, editor of the *Mercury*, agreed. Lanney's body should have been allowed to lie in the grave and then, in the course of time, raised so the bones could be preserved. Alternatively, the colony's men of science could have used their skills to embalm the body.

All those engaged in making public comment about this event turned to the public to support their view. And as Stefan Petrow has shown, colonial politics played a large role in the affair.[20] One of Crowther's political rivals, John Davies, was pleased to see strong public indignation at the mutilation of Lanney's body, which he said in the *Mercury* (9 March 1869) showed

that the tone of public feeling is sound, and that the common people have a better appreciation of decency and propriety than such of the so called upper classes and men of education who could have committed a flagrant outrage from which the greatest ruffians in Tasmania's worst day would have shrunk.

Muggs judged the new Tasmanians more harshly. He believed that the British colonisation of Van Diemen's Land had degraded both coloniser and colonised, for in introducing civilisation to '7,000 heathens', British Tasmanians had become 'murderers, and something worse'. William Lanney, he said, was a 'cruelly dishonored fellow-creature', and in mutilating his body, the colonisers had exposed themselves as being 'even more revolting than cannibals'.[21] This reminded everyone of European theories about racial degeneration, which was a fate that could affect any race of human beings.

Tasmanians of British descent, many of whom felt doubly tarnished by criminal roots and native extinction, understood how easily they might become degenerates in European eyes. As 'An Old Colonist Too' expressed it, the 'stigma for which we are already too notorious' had been multiplied by the mutilation of Lanney's body, and he worried that Crowther's subsequent victory in the Legislative Council elections revealed that Tasmanians, as a people, were immune to shame.[22]

The editor of the *Cornwall Chronicle* was anxious about the degree to which the government had connived at Lanney's mutilation, and the effect this would have on Tasmanians' 'national character'. History, he thought, avouched for the fact that

> no large section of any community, much less a Government, ever acted on indecent or immoral principles without inducing a proportional degradation in the virtue and dignity of the individuals who form the mass of the people.[23]

There was a strong concern in all of this that the world was watching and judging Tasmanians' actions. 'An Old Colonist' called for a response that would

> show to the world that the colonists, as a body, are indignant that so foul a deed should have been perpetrated; and thereby, in a measure, wipe out the stain that has been inflicted upon the colony by a few individuals.[24]

The editor of the *Tasmanian Times* also denied that the stain was born by the public in general, charging that it rightly belonged to the perpetrators of the act, who had inflicted 'infinite damage on the credit of the colony'.[25] The medical profession, the legal profession and the executive ministers of the law had all been represented in an act of 'criminal indecency ... upon a buried Christian corpse', which had 'outraged the public sentiment of the community, broken the law of the land, and brought discredit on Tasmania at home and abroad'.[26]

These concerns were very worrying for a society that had pondered for decades about how best to raise the tone of a population bearing such morally impoverished beginnings as did many British Tasmanians. However, just which part of the mutilation of Lanney's body was considered to be most reprehensible was somewhat unclear. While John Davies predictably directed his ire at the decapitation, the editor of the *Tasmanian Times* thought this part of the action involved no moral turpitude, while snatching Lanney's body from the burying ground did. The *Times* had located a witness to Lanney's resurrection who named George Stokell as one of the men involved.

Sordid rumours were also circulating in Hobart Town about indecencies having been perpetrated by Stokell on Lanney's body. While filleting on behalf of science, the resident surgeon had taken the opportunity to souvenir a portion of Lanney's skin for a friend. According to a statement later made by the hospital's gatekeeper, Charles Williams, Stokell had a tobacco pouch made out of part of the skin. Stokell's friend remembered that the doctor 'skinned the body and gave me a piece sufficient to make a handbag, and a very excellent soft bag it made'.[27]

Such activities were not unique to this dissection. Pieces had also been taken from other famous bodies, as when the University of Edinburgh's Professor of Anatomy, Alexander Monro Tertius, dissected the murderer William Burke.

But contrast Stokell's work on Lanney with how such things were undertaken by men who really valued scientific knowledge. When William Flower read the Tasmanian newspapers Crowther forwarded to him, how horrified he would have been. That crude harvesting of bones and the stealing of pieces of skin to make pouches and bags was an illustration of vital opportunities lost to comparative anatomy. Two years before Stokell was cutting so indiscriminately, thinking that only the bones were of scientific value, Flower and James Murie had dissected a young 'bushwoman' at the Royal College of Surgeons, and their observing eyes missed nothing in this careful process, which was recorded. They laid her on a table, then carefully walked around the body, closely observing the surface anatomy. Every feature of the face was recorded. A curl was straightened to assess its length. They measured her arm, resting down her side, then the breadth of her palm and the shape of her fingernails. They noticed how European shoes had deformed her feet. An outline of her body was traced onto a sheet of paper for later comparison with the beautiful female figure given in G.V. Carus's *Proportionslehre*. Only then did Flower take up his scalpel to cut. Every cavity and system was thoroughly investigated and recorded, before parts of her were preserved and the rest disposed of, except for her skeleton.[28]

Whatever might be said of George Stokell's surgical skills, he was no comparative anatomist. And neither were the Royal Society men on whose behalf he worked.

In the Name of Science

In public reflections on all this work, science appeared as a strange, even warped activity in Tasmania. One letter to the editor of the *Tasmanian Times* viewed the scientists with contempt, as 'the expectants of the ghastly relics [who] had just calculated upon obtaining their object by some secret and stealthy means of which the public should know nothing'.[29] Even John Davies, who believed the Royal society's Fellows were those best entitled to hold Lanney's body in the interests of the public, wondered a little about the Fellows' scientific competence. He found it a singular circumstance that no-one had thought before now to secure a perfect skeleton of a male Tasmanian Aboriginal for the museum. This was an opinion that William Crowther shared. Writing to Flower, he said that although the society had existed for twenty years, it had never thought of procuring a skeleton until he had moved on Flower's behalf. He called the society's members 'a mere clique of *would-be savant*[*s*]'.[30]

The *Mercury* soon reverted to form in suggesting that Crowther was little more than a mercenary in the cause of science, in contrast to the legitimately scientific desires of the Royal Society. The editor darkly suggested that Crowther had stolen the skull for financial reward rather than for science. This differentiation between 'pure' science, as practised by the Royal Society, and *ersatz* science, as practised by Crowther, was an attempt to restore some prestige to the local society and its scientists.

However, some members of the public saw all of these mutilations as a crime and were not fooled by attempts to differentiate between the perpetrators. Job Muggs dismissed the claims of the Royal Society's Fellows and their museum, and of Crowther, in rhetorical flourishes that cut to the bone. He asked whether it was 'Science or morbid Curiosity ... that makes this strange and untimely demand?' and suggested

> [I]f it is curiosity ... he being the last black man—why not gratify further the insatiable appetite for strange sights with which [the museum] ... is known to be afflicted. Why not ... place in the same gallery the skeleton of the first White man who murdered a Black one and ... as a lasting memorial of the Tasmanian workers in the Royal cause of "Science," ... place by the headless skeleton of "King Billy," that of the "diplomaed" butcher and lawless ghoul who mangled and stole one portion of the remains after death, and in the darkness of night robbed the churchyard of the other. Curiosity calls for such food, and its appetite ought to be appeased.[31]

Muggs's criticism extended to individual Fellows of the society, and he called James Agnew the 'medical magnate' who ruled the roost there.

Agnew had been a strong and persistent advocate for government assistance for the society's museum, whose new premises had largely been funded with public money. He was also the man to whom George Stokell had turned to report the theft of Lanney's skull. When Agnew and Allport instructed Stokell—who was a hospital employee—to remove the feet and hands, the resident surgeon did so unhesitatingly. Agnew sought to justify this in a letter to the press that restated evidence Stokell had already given to the men on the government's Board of Enquiry —who had been startled to learn the extent of the Royal Society's power within the hospital.

Everyone knew the value attached to a perfect skeleton. A skeleton without its feet and hands was comparatively useless. But if perfection could not be attained, the skull was the next best thing.

Not surprisingly, Agnew chose the anti-Crowther *Mercury* in which to respond to criticisms being made of the Royal Society. He defended the society's actions and Stokell's in terms that suggest he was arrogantly out of touch with much public feeling on the issue:

> It will, I am certain, be admitted by every enlightened individual that the action taken (unanimously) by the Council of the Royal Society was such as they were bound to do. The skeleton of our last aborigine ought unquestionably to have been reserved to all futurity in a Tasmanian museum ... As a native of Tasmania he [Dr Stokell] felt his first duty was clearly due to the land of his birth, and this he has faithfully performed, uninfluenced by more personal considerations.[32]

Agnew hoped that after reading his letter, the public would see that Crowther was the real culprit. But this sanguine expectation was not shared by others. On finding several months after Lanney's death that the Royal Society had requested the Council of the Royal College of Surgeons in London to return Lanney's skull, the editor of the *Cornwall Chronicle* poured ridicule on the society's pretensions.

> Our local *savans* must have an extraordinary idea of their own importance, or a very poor opinion of the shrewdness of the eminent men who compose the Council of the Royal College, if they entertained the idea that an application of this kind would forward their endeavours to add the cranium of Lanney to the mutilated trunk they abstracted from the burrying [*sic*] ground in Davey-street.[33]

That the local Royal Society failed to understand their collection's relative unimportance in the British scheme of things was a view that fitted well with Crowther's. Stokell had informed the Board of Enquiry that when he told Crowther the Royal Society had the best right to Lanney's body, the older man had responded that Stokell was 'a fool to keep it in a paltry place like Tasmania'.[34]

These words were meant to describe the state of the local Royal Society and its museum, as much as the colony itself, for the Tasmanian Royal Society was nothing like the prestigious body that bore that name in London.

In the Name of Medicine

Woven through the criticisms of the state of Tasmanian science lay deep concerns about the practice of medicine in the colony, for the Lanney affair had exposed the General Hospital as an institution in which those who died were routinely 'mutilated' by medical men.

Soon after the discovery of Thomas Ross's skull in the burying ground, the gossip that filled Hobart Town was fuelled by an increasing realisation that anybody in Tasmania could suffer Ross's post-mortem fate, given a descent into poverty and the lack of friends sufficiently assertive and well-to-do to claim their body for burial. Hundreds of people visited the graveyard in the hope of inspecting the evidence for themselves, making the kind of public furore that medical men had long sought to avoid. Dissecting was generally only spoken about when news of work on bodies escaped the places that normally contained it, as had most infamously been the case when Robert Knox was found to have purchased Burke and Hare's victims. Knox rationalised his subsequent decision to remain silent rather than defend himself in terms all medical men would understand when he said that disclosures about what happened in dissecting rooms 'must always shock the public and be hurtful to science'.[35]

The passing of Britain's Anatomy Act had followed hot on the heels of several London episodes of burking, but as none of the antipodean British colonies replicated that Act, in Tasmania in 1869 the only bodies legally available for dissection remained those of murderers. Yet here was news about just how readily other people became subjects for dissection in this colony.

As we have seen, the first brief ministerial enquiry had had the effect of making the mutilation of the dead in Hobart Town's hospital appear to be a Crowther family idiosyncrasy. The second enquiry achieved something else. It made it obvious to all that such activities were, instead, a matter of common daily practice at the hospital. Even the *Mercury* could not ignore the fact that medical delinquencies had been exposed. Its editorials suggested that the cure was to be found by placing more power in the hands of the resident surgeon, for it was apparent that somebody needed to keep a 'check on the treatment of ... dead bodies' in the institution.

John Davies thought through the implications of making bodies so readily available to medical men, asking

> what security has the Hospital patient labouring under some compli-
> cated or curious disease, that his case may not prove too interesting to
> be properly treated in the hurry to have the mystery solved on the dis-
> secting table?

No medical man attached to the General Hospital—even Crowther's enemies—would have felt comfortable with this implication.

Distinctions were made between what had happened to Lanney and the 'barbarous conduct' on the body of the unnamed white man. The 'fact that another body as well as that of the aboriginal was mutilated, shows that there is no respect of persons', and that the hospital's honorary medical officers seemed to have a 'license to hack and hew the bodies of those who die in the Hospital for their own amusement, or the instruction of their pupils'.

What had happened to Ross and Lanney had also exposed 'looseness in the management of one of our charitable institutions' that could

> shake public confidence in the Hospital, and lead to shutting the doors.
> People will endure any suffering and sickness rather than seek relief
> where, if death closes their eyes, their bodies may be treated with an
> indignity from which, applied to a dog, a sensitive mind would revolt.

As matters stood, Davies said 'the occupation of a bed in the Hospital is but an introduction to the scalping knife'.[36]

The *Tasmanian Times* began to refer to George Stokell as 'the resurrection man of the worshipful society of body-snatchers'.[37] This was precisely the kind of language that had brought medical men into such ill-repute in Britain prior to 1832.

Crowther and his supporters took the opportunity to shine critical attention on the composition of the hospital board. Crowther illustrated the degree of the government's power over a board that only claimed to be independent. It was a point he and others had been making for years, and he was not about to let this

Caricature featuring Dr William Crowther in St David's
Cemetery, date unknown, artist unknown. This confused image is
more redolent of earlier nineteenth-century images of grave
robbing than of the 1869 Tasmanian resurrection. On the left, the
body being hauled out of the grave by a rope around its neck looks
nothing like William Lanney. However, the two ruffians in the
bottom right-hand corner may represent the hospital's gardener
and messenger who assisted George Stokell in this night's work,
and the fight between two groups of resurrectionists may refer to
rumours that William Crowther also planned to steal Lanney
from his grave that night. The word 'Sacred' stands out on one of
the gravestones. This sketch places resurrection as work overseen
by the devil. Courtesy of the W. L. Crowther Library,
State Library of Tasmania.

new opportunity pass. Dry had been careless when writing to inform Crowther of his suspension from the hospital. He said that he had issued 'instructions' to the board. Crowther immediately wrote back to inform the Colonial Secretary that 'I do not hold my appointment as an Honorary Medical Officer of the General Hospital from the Executive Government, but from the Board of Management to whom I am alone responsible'.[38]

Weeks later he wrote another furious letter to Dry—nib digging deeply into paper, words underlined to increase their effect—to make the Colonial Secretary understand how deeply he had overstepped his authority in interfering in hospital business. The government, Crowther pointed out, did not own William Lanney's body and had no power to dispose of it.

Some charged that Hobart Town's hospital board was a 'little cosmogony' (its numbers had dropped from twelve to five) in which '[j]ealousy, intrigue, corruption, and immorality [have] so long been in the ascendant, [that] nearly every spark of manly feeling has been extinguished'. The *Times* suggested that unless improvements were made, Tasmanians should 'admit at once that Board Management is beyond our moral capacity and intellectual reach, and revert without delay to the old system of official management'—the system, that is, which had been suitable for a penal colony.[39]

In the Name of Christianity

Science and medicine were not all that was found wanting by this episode. Critical eyes also focused on how Christianity was practised in the colony with regard to Christian burial. Much public comment related to notions of the 'sanctity' of a grave, in the light of what seems to have been the first revelation of the practice of graveyard body-snatching by medical men in Tasmania, in stark contrast to the situation in Britain prior to 1832.

Lanney's funeral had been seen by some Tasmanian settler-colonists as an opportunity to pay their respects to a passing race.

> There are many amongst us, probably, who will view with regret, and
> perhaps with compunction, the extirpation of the original occupants of
> the soil of Tasmania. As we can now no longer provide in any way for

the comfort and protection of these unhappy victims of advancing civilization, it is only decent that we should honour their memory in the only way in which it is now possible to exhibit our appreciation of their merits, and our sorrow for their extinction.[40]

However, by the time readers of the *Tasmanian Times* had digested these words with their breakfast, Hobart Town was already filling with rumours that Lanney's body had been tampered with overnight. Having heard the gossip, on arrival at the General Hospital to accompany his coffin to the church, Lanney's whaling friends demanded the lid be opened so they could see the body for themselves. John Graves, who had sent out the funeral invitations and also walked away with Lanney's hands and feet in his pockets, hovered protectively over the coffin as its lid was lifted. What Lanney's friends saw was not recorded, but it was sufficient for them to insist that the coffin be sealed immediately. A decision seems to have been taken that nothing must be allowed to interfere with the public funeral (though afterwards several of these men wrote to the Colonial Secretary to demand an official enquiry). The coffin was sealed with an old brass stamp found in the hospital dispensary, which left the word 'world' in the wax. Then it was covered with an opossum-skin rug, 'fit emblem of the now extinct race', on top of which were laid

a couple of native spears and waddies, round which were twined the ample folds of a Union Jack, specially provided by the shipmates of the deceased. It was then mounted upon the shoulders of four white native lads, part of the crew of the Runnymede ... The pall was borne by Captain Hill, of the Runnymede, himself a native of Tasmania, and by three colored seamen, John Bull, a native of the Sandwich Islands, Henry Whalley, a half-caste native of Kangaroo Island, S.A., and Alexander Davidson, an American.[41]

Led by this profusion of 'natives', white and black, the funeral procession grew as it moved along the streets between hospital and church. Eventually more than 100 people followed the coffin, and after a ceremony in St David's church, Lanney's remains received a Christian burial, though during his life (despite the best efforts of the Queen's Orphan School and George Augustus Robinson) he had shown few signs of religious allegiance.

Those Christian gentlemen in attendance at this ceremony understood death to be more a point of departure that held the possibility of life everlasting than an end in itself. British ways of understanding and marking death and burial were part of the cultural baggage they had brought with them to the antipodes. For many, graveyards were sanctified places of rest where bodies lay awaiting the final day of resurrection. It is difficult to understand how men like George Stokell, who was present at Lanney's funeral and planning a midnight resurrection of his body, rationalised their participation at church and graveyard. Did they see this ceremony as a politically necessary farce, and understand 'King Billy' as a heathen whose body was of little Christian significance? Even so, what they subsequently did in that graveyard also offended a secular sense of what was and was not 'decent' in the treatment of human remains.

And what of John Graves, who thought of himself as a true 'friend' of the Aborigines, and even named his own children after them? He knew the Aboriginal Tasmanians cared for their dead, and that the desecration of burial sites was abhorrent to both cultures.

Reverend Cox, who presided over Lanney's funeral, later found his role in the proceedings questioned. A newspaper correspondent using the name 'Admirer' asked of whom Cox had been speaking in intoning the words 'the soul of our dear brother here departed'. Was it 'the Presbyterian whose head he was burying? or of Lanney'?[42] Job Muggs saw Lanney's burial as a Christian man as nothing more than a form of blasphemy that further illustrated the hypocrisy of British colonialism in Tasmania:

> [U]nder the pretence of Christianising and civilising [the Aborigines], [we] debauch and degrade them—next exterminate, and finally in defiance of the practices of even barbarous and civilized peoples, ancient or modern, insult and mangle their dead bodies after the manner of ghouls and jackalls—and, worse than all, in derisive mockery of the laws of religion ... What a comment upon religion and civilisation! ... the whole of the scandalous proceedings are viewed with comparative

indifference by those who most affect religion—and are merely laughed at by that large portion of society which does still, and has always regarded the native races with far less consideration than they are accustomed to bestow upon their dogs or cattle.[43]

For many Tasmanian settler-colonists, participating in Lanney's funeral was more an opportunity for play than for mourning the passing of a man or his race. Traces of such feelings are evident in the publication of irreverent ditties. People's expectation of being entertained, and having their curiosity satisfied, is also revealed by the fact that visitor numbers to the Royal Society's museum doubled in 1869, presumably in the belief that Lanney's headless skeleton would be on display.

The truth is, what had happened to Lanney's body provided both an opportunity for serious reflection *and* a temptation to laugh. Even Job Muggs said that when he had initially sat down to write to the *Cornwall Chronicle*, he had intended to treat the subject with less gravity than he actually did.

TRACKING WILLIAM LANNEY

Of course, there was more to the story than all this local bother. After all, William Crowther had taken Lanney's skull to fulfil a promise made to the Conservator of a British museum. With hindsight, we know this episode also led to Morton Allport's solo career as Europe's premier local collector of Tasmanian bones. In the long term, the events of 1869 came to define William Crowther's place in Tasmanian history. Nothing else in his eventful life is as well remembered today, despite the statue erected to his memory in Franklin Square after his death, which was paid for by public subscription and has him standing tall. Instead, this one episode of dissecting has come to define him. In histories he is presented, at best, as an arrogant braggart and, at worst, as a heartless racist.

Meanwhile, Morton Allport—who suggested the further mutilation of Lanney's body, oversaw the robbing of his grave, and arranged other thefts from graves on Flinders Island and at Oyster Cove—has entirely escaped such judgements. Let us dig deeper into this mystery.

Soon after Crowther failed to procure Lanney's skeleton for the Hunterian Museum, he wrote to Flower to inform him of what had happened, and advise that he had sent a couple of men to open up half a dozen Aboriginal graves at Oyster Cove to obtain a skeleton for Flower, now that all official channels for doing so were closed. However, he warned Flower not to be too hopeful, for Sir William Denison, an ex-governor of Van Diemen's Land, had sent down for three heads while he was in India, and these had been taken from Oyster Cove.

A month later, Crowther reported to Flower that each grave his men broached was found to contain only a headless, though well-preserved trunk. But this was not true. There were complete bodies in the Oyster Cove graveyard, and within a year or two, Morton Allport would obtain Bessy Clark's skeleton from there. Forty years later, Crowther's own grandson would dig up several more.

So what are we to make of Crowther's letter to Flower? It may be that the men he sent to search these graves had no taste for the job. Or that he sent nobody at all. He had also promised Flower that he would lose no time in having the graves on Flinders Island investigated. Then he let the matter slip entirely, for there is no more talk of Tasmanian bones in Crowther's letters to Flower.

The bodies that bound Crowther to Flower were those of whales and sons, not Aboriginal Tasmanians. Soon after the Lanney affair, Crowther sent young Bingham to England to continue the medical education he had only just begun under his father's tutelage. Crowther probably did this to remove a sensitive boy from the scandal of his involvement in the Lanney affair. Bingham was quite unlike his confident father and his older brothers, and Crowther knew he was unprepared for England, where his education was concerned. He was weak in the Arts, at a time when the Arts were part of a physician's education. Crowther blamed this on Bingham's bouts of ill-health, which his father said had compelled him to put him at once to Anatomy and hospital work. This ill-health was a matter of nerves rather than a lack of physical stamina.

In introducing his son to Flower, Crowther sought at first to give him the kind of hardy character and proclivities he desperately wished him to have. He wrote that Bingham was in the process of dissecting the body of 'a Native from "Savage Island" in the South Pacific', a whaler who had died of pneumonia in Tasmania. He said the boy was sending this body to the Hunterian Museum and emphasised that Bingham wished it to be considered as a donation *from himself*, which gave his father no little pleasure.[44]

But the fact is that Bingham was not like his father. He was a cause for family concern. Crowther admitted to Flower that he was a little anxious about him. This son, he said, had a different disposition and temperament from the others, being of gentlemanly feeling and manners. He asked Flower to use his influence at the College to have Bingham's local medical training recognised, though this meant finding a way to work around the institution's rules. He did not wish Bingham to stay in London for four years.

Crowther rationalised the request as being a matter of family economics. But later letters reveal he was also concerned about Bingham being so far from home. London, he thought, was in truth a dangerous place if one were left alone, especially Bingham, with his sensitive temperament.

Most of Crowther's words to Flower over the years were penned to maintain his connection to a metropolitan world of science of which he greatly desired to be a part. He had already sent to 'the Old Country' the precious head of a killer whale, and the Zoological Society of London had recognised his work on the cetaceans by making him a Corresponding Member. Crowther told Flower with pride this was in 'High recognition of my Services to Zoological Science in collecting the Tasmanian Cetacean' and placed it as an unsought compliment, saying he believed 'it is unnecessary to further thank or reward a man for doing that which he believes to be his duty'.[45] In his next letter, he welcomed Flower's news that he, Crowther, was about to receive the Royal College of Surgeons' prestigious Gold Medal, of which there were only five recipients in the world. He said that 'to a person engaged in the cause of science', it was 'always satisfying to find that his labour has been appreciated more particularly so when everything in this utilitarian age is accepted as mere matter of fact'.[46]

Such self-perceived nobility of purpose might make us cringe, but it pales into insignificance when compared to Morton Allport's letters about the rewards he received for his engagement in science. They contained lists of the important men he counted among his 'intimate personal friends', the organisations that had honoured him, and the wonders contained in his beautiful cabinets. In one brief moment of reflection about how these letters might be read by those to whom he sent them, he said he hoped not to be thought 'an awful snob', but that he could not help it.[47] In another letter he boasted of holding up his end against the English 'swells' with whom he competed for prizes in photography.

And Allport had other ways of holding up his end against some of his fellow colonists, such as William Crowther. He decried 'the class of colonists who [now] fill the old places', who are 'not to my liking'.[48] In another letter, he spoke of attending an engagement at which Crowther's daughter, Caroline, was present, and overhearing an old colonist 'addressed by the Elegant and accomplished daughter of the proprietor of "Bird Island"' in the following way: '"Dear me Mr Terry your moustache doesn't seem to get on very well, never mind I'll bring you a … pocketful of pa's *Guano*"!!'[49]

If we can be sure of anything about Morton Allport, it is that winning mattered quite as much to him as it did to William Crowther. It was another context in which their competitive quest to possess William Lanney took place. In 1871, when all enquiries had failed to locate Lanney's skull, Allport set about the business in another way. As we have seen, he had those two Tasmanian skeletons dug from their graves and shipped them, unsolicited, to Crowther's correspondent, William Flower, together with a letter offering a third skeleton if only the Royal College would return Lanney's skull to the colony. This valuable gift—a native pair of the kind Flower had asked Crowther to procure—was loaded with sub-texts. It is difficult to imagine a more insulting gesture from Allport to Crowther. It also shows how much power Allport wielded within the Royal Society, whose museum he now curated, and whose Fellows had belatedly come to realise the value of Tasmanian skeletons.

How important the skull of the last man must have seemed to them, that they would give up a perfect pair—and, Allport promised, further treasures—for its return: 'I would willingly give another perfect skeleton for [Lanney's skull] as the rest of the bones are in the Tas. Roy. Soc. Museum and could thus be considerably increased in value without any detriment to the English Museum'.[50] These men were obsessed with gaining possession of the last man. But there was also something very personal in this determination.

Allport had set off in this direction after meeting a man named Basil Field, an acquaintance of Flower's, who had been visiting Tasmania. In London, Field lived

in Lincoln's Inn Fields, close to the Royal College of Surgeons. Allport was fol-
lowing Field's suggestion in offering this museum exchange to Flower. Five months
later, when Field had returned to England and made some enquiries there on
Allport's behalf, he received a letter of thanks from Allport for his trouble. Allport
said he had written to Flower and was expecting his answer by the next mail.
When it arrived, he told Field that it confirmed what he had said about the skull.
The letter to which he referred has not been preserved, but it must have informed
Allport that the College did not possess the skull.

Next Allport made a strange request of Field:

> I now learn that young Crowther (a pupil at Guy's) has the skull which
> his father stole from the General Hospital here. If I was myself in
> London I would soon have the skull back by some means but as I am
> not could you devise any method of obtaining it?
>
> I would willingly give £10 to recover it and the young fellow daren't
> make any disturbance if he loses it as he well knows it does not belong
> to him. Would you think this over and drop me a line[?][51]

It seems that Field made no response to this decidedly odd letter. Allport was
asking him to hire or bribe somebody disreputable to locate Bingham Crowther's
rooms and rifle through them for the skull. Perhaps Field did not share Allport's
enthusiasm for stealing bones.

Soon Allport was writing to the greatest of all European skull collectors,
Joseph Barnard Davis, and Lanney's skull was still on his mind. He told Davis that
the skull was in Bingham's possession. But something had happened during the six
months that passed between this letter and the earlier one to Field (July 1872 and
January 1873), and it changed the tenor of Allport's importuning. He now called
Lanney 'the (so called) last male Aborigine'. 'So called' he said, because 'I have
strong grounds for suspecting that Lanney was a half-caste and the form of the
head, as seen in the photographs, is utterly unlike the true Tasmanian'.[52]

What are we to make of this? Allport was planting the idea in English scientific minds that the skull Crowther had taken so much trouble to obtain was,
scientifically speaking, quite worthless. Thwarted, despite considerable effort, in
obtaining Lanney's skull, he may have set about making it insignificant in scientific

eyes. The photographs to which he referred had been taken by Charles Woolley in 1866 for Melbourne's Intercolonial Exhibition, and they were widely available from that time. It is inconceivable that Allport had not seen them before now, for he had acted as one of the Tasmanian Commissioners for the Exhibition, at which both he and Woolley had won medals for their photography.

Allport was also, still, intensely interested in locating Lanney's skull—no matter that the reason he now gave was a different one. He told Joseph Barnard Davis, a man who had no interest at all in the skulls of 'half-caste' people, that he should travel to London to visit Bingham Crowther at Guy's Hospital to see the skull for himself. He said that so long as Davis was careful to conceal the fact that he had obtained the information from Allport, he thought the doctor would be able to examine the skull and share with his colonial correspondent his opinion as to its 'genuineness'.[53] As to its presence there, more like.

William Lanne, Coal River Tribe, 1866, by Charles Woolley.
Courtesy of the Tasmanian Museum and Art Gallery.

Or were there genuine doubts about Lanney's race? No letters exist to indicate that Davis followed Allport's advice by hounding Bingham in this way. When William Crowther heard what Allport had done, he wrote to Flower and planted a doubt of his own about the authenticity of one of the skeletons Allport had sent to the Royal College. There was some doubt as to its 'character', he said, since when he had had the Oyster Cove graves examined 'with the exception of the "half-castes", all were *headless*, and no burials have taken place since'.[54]

Had Davis come up to London to visit Bingham he would, in any event, have been wasting his time, for Allport was wrong in believing Lanney's skull was in Bingham's possession. It still lay in William Crowther's home. This may seem odd, since Crowther had had many opportunities to forward it on over the years, if not in the company of his sensitive son, then in one of his own vessels, or as baggage in another, as had been the case with the skeletons of whales. Instead, he kept it. Despite his bombast, Crowther may have regretted his actions that night in the hospital's dead house. He was an impetuous man, and he had not foreseen the trouble they would cause him, which included the unwelcome experience of being pilloried in the press by an ex-convict, John Davies, of the *Mercury*.

The attacks on him were, he said, 'the most violent political attacks that have ever been made on any private individual within the Australian Colonies' and, although he was 'not one of a class devoid of British Pluck', he felt their effects. Referring to the Gold Medal he was about to receive from the Royal College, he said that 'if the College give me any such and as many Fellowships I should not be compensated for the annoyance I have had from the Skull of the last Aboriginal'.[55] Though he thought he was a match for the continued attacks against him, and had intended to charge Richard Dry with lending his official position for the purpose of damaging his reputation, he felt his hands were tied by Dry's sudden death a few months after the Lanney affair.

Years later, Crowther sought to interpret the Lanney affair as nothing more than a good joke. But these were the words of a man who knew that if the whole thing were a joke, then he was its butt. Despite all this, Crowther's claims to being a man of science must be taken seriously. He understood the contemporary argument—even when it was made by his enemies—that Lanney's skeleton must be complete to be of most value to science. He also believed that the bones of the last man were a necessary addition to the local collection. This is evidenced by his early query to Flower about the possibility, should the Hunterian Museum obtain the whole skeleton in the future, of making a cast of it and sending it to the Royal Society (Flower replied that it would not be possible). Meanwhile, he worked behind the scenes with an unnamed 'friend of all parties' to reunite the skeleton, preferably in London.

After Allport's claim that Lanney was a half-caste, Crowther reassured Flower that once the skull reached him safely, he would have not only a true one, but the last of the race. Through his unnamed friend, he worked to persuade the Royal Society to ship the rest of Lanney's skeleton to the College. But it never happened. Instead, the society's Fellows built on the rumour that Lanney's skeleton was nothing worthy of note.

Years later, James Agnew delivered a paper on 'the last Tasmanians' to the 1888 Sydney meeting of the Australasian Association for the Advancement of Science. Against the view of men such as Crowther, who saw in Tasmanian skulls the possibilities for improving the race, Agnew insisted that Lanney's skull illustrated no such thing. He said it had erroneously been sent to the Royal College of Surgeons as being a pure Tasmanian Aboriginal, when all the time it was the skull of a man who had the blood of a New South Wales Aborigine in his veins. The gift to the Royal College, Agnew lamented, had been an unfortunate presentation.

Lest any of his listeners see his charge as being an act of continuing spite against Crowther, he added 'I only refer to this matter at present, because … the scandal attached to the removal of the head [was] a source of great trouble to Truganini … [whose] grief after her husband's death was something terrible'.[56]

You would hardly think this was the same man who badgered the government for two years after Truganini's death in an ultimately successful effort to obtain her bones, would you?

FALLOUT

Though due to the Lanney affair William Crowther had lost his position as an honorary at the hospital, in the months following the discovery of Thomas Ross's skull he experienced many moments of personal and professional satisfaction. The first was in being elected to represent Hobart Town in Tasmania's Legislative Council. His candidacy did not suffer at all from his theft of Lanney's skull. Crowther was a man of large ideas, and a popular candidate. At the public meetings his supporters staged, he successfully presented himself as a man of the people who had enemies around him in 'great force'. He minimised the scandal by calling it a hubbub and hoping it would not distract Tasmanians from the 'great political interests' at stake in an election in which he stood for equalising the distribution of land, extending the franchise, making the taxation system fairer, and reforming the method of parliamentary representation. Without these progressive reforms, he argued, the colony could not hope to prosper.

It was an articulate rallying cry which made those penning anonymous letters to the press to remind voters about the Lanney affair appear small-minded in comparison. Public speeches were also made decrying the way Crowther had been treated by the government. Twelve hundred signatures were presented on a petition to Governor Charles du Cane demanding his reinstatement at the General Hospital. While the petition failed, Crowther was elected to the upper house of parliament with an impressive majority, despite the desperate machinations of a group of men led by Allport and Agnew, which included dragooning a reluctant alternative candidate of their own into the political contest.

Other satisfactions came in seeing off both George Stokell and Richard Dry. Stokell lost his position as resident surgeon at the hospital when the hospital board changed the resident surgeon's role and reappointed George Turnley, a more senior man, in his stead. Over the next few years, when hospital Chairman Alfred Kennerley had become Colonial Secretary and Crowther was a member of his ministry, several parliamentary enquiries were established to investigate the way the General Hospital was run. Crowther was in the thick of them, and he took vengeance where he could. As we have seen, Turnley already had reason to dislike Crowther, who had treated him as a nonentity in the hospital. But Turnley was undone when these enquiries produced reports that were highly critical of the standard of care provided in the hospital.

Two of Crowther's sons—Edward and Bingham, who had returned after completing their educations in England—gave evidence before the first enquiry in 1875. Edward had made a surprise visit to the General Hospital and pointed out its deficiencies in public. Compared to British hospitals, he charged, Hobart's General Hospital did not separate patients with infectious diseases from 'clean' patients, nor the dying from those who would recover. He spoke of being horrified to discover the contents of some of the 'cells' in which patients were housed, which were furnished with 'a mattress on the floor, a leather bucket for a urinal ... [and] a human being'. There was no hot water in the bathroom. The nurses' quarters were inadequate. And hospital steward Charles Seager's ducks were running riot in the grounds. According to another critic, excrement also slid remorselessly from the outhouse towards the main hospital building.[57]

Turnley and Seager tried to bluff their way out of these charges by denying everything. The steward, who was responsible for overseeing the hospital's cleanliness, said defensively that the 'wards are generally cleaned once a week, sometimes once a fortnight'. Two years later, he was removed from his position at the hospital when a second enquiry heard of his insolence towards the new Lady Superintendent of the nursing staff, the redoubtable Miss Abbott.

The final insult to George Turnley came in the form of William Crowther's reinstatement to his position of honorary medical officer at the hospital in January 1877. Turnley resigned and, in a final triumph, Bingham Crowther was appointed in his stead. It may have taken eight years to achieve, but this was sweet revenge for the Crowthers.

As for Richard Dry, he died suddenly in August 1869. This meant that when an Anatomy Bill belatedly passed through the Tasmanian Parliament later that year, Dry was not among those to deliberate upon it, though other men who had been intimately involved in the Lanney affair were. Crowther sat in the Legislative Council, together with Hobart Town's ex-mayor James Milne (who was now the Colonial Secretary), and Alfred Kennerley. These men debated the terms of the Bill, which took the British Anatomy Act as its primary reference point, and liberalised it. In Tasmania, from this point on, 'anatomical examinations' were only to be carried out in licensed premises, by licensed practitioners and their students. As in Britain, the Tasmanian *Act for regulating the Practice of Anatomy, 1869*, made it legal for people to donate their bodies for anatomical examination. Others whose bodies could be 'examined' were people not claimed for burial by family members when they died. The Tasmanian Anatomy Act also replicated the earlier British

Act in other ways. It contained no clause requiring officials to alert family members to a person's death, nor a means to forewarn the public that dissection might be their post-mortem fate. And while the British Act gave relatives forty-eight hours to claim a body for burial, in Tasmania this was reduced to a parsimonious twenty-four.

Had the Tasmanian Anatomy Act been in place six months earlier, the result for Thomas Ross and William Lanney would have been no different. Ross had died in a hospital in a colony 12,000 miles from his home. He was unmarried, and no relatives came to claim his body. Friends may have wished to do so, but friends counted for nothing under the new Anatomy Act. So after his death, Ross would still have been carried to the hospital's dissecting room. As for William Lanney, though he was understood to have a living relative, in the form of a wife, Truganini, whether she was legally recognised as such is another matter. Besides, she lived at Oyster Cove, in the care of Morton Allport's acquaintance Superintendent Dandridge, and so may not have learned of Lanney's death within the requisite time.

The whereabouts of Lanney's skull remains a mystery. This has allowed some imaginative stories to be told. Like longed-for sightings of the extinct Tasmanian thylacene, or tiger, there have been some exciting moments of 'discovery', though so far all have led to disappointment. Most commentators believe that Crowther sent the skull to England. During the 1870s, Thomas Smart, one of his fellow honoraries at the General Hospital, believed the skull had arrived in England inside the body of an 'islander', which gave it a 'proper setting' for the voyage.[58] Could William Crowther have been so cruel as to make Bingham prepare the body of that man from Savage Island to receive Lanney's skull? No, for the dates are wrong. He still held Lanney's skull in his home when this Pacific Island man travelled posthumously to London. In a letter to Flower of 19 April 1873, Crowther said he had heard that the Royal Society of Tasmania was going to send Flower Lanney's skeleton and if this proved true, he would send on the skull. But as it turned out, the Royal Society kept Lanney's skeleton.

There are also tales of Crowther shipping the skull to England so poorly prepared that it began to stink and some sailors threw it overboard. From what we know of Crowther's skills and experience in preparing skeletal material to send to Britain, this seems unlikely.

The most probable sighting of the skull was in Edinburgh. Upon William Crowther's death in 1886, his son Edward found several skulls in the family home, one of which was thought to be Lanney's. Edward gave this skull to a friend, who in turn took it to his ex-teacher, Professor William Turner, in the Department of Anatomy at the University of Edinburgh. Turner was a comparative anatomist with a fine collection of skeletal material. It included the articulated skeleton of William Burke, still on display to serve the post-mortem part of his sentence for murdering all those people in 1827–28.

But when Edinburgh University repatriated its collection of Tasmanian bones in 1991, a photographic imposition revealed that Lanney's was not among them at all.

CHAPTER SIX
THE USE AND ABUSE OF
HUMAN REMAINS

WHERE DID MARY MCLAUCHLAN, WILLIAM LANNEY and Thomas Ross finally come to lie when they left that liminal space of the dissecting room? And what became of Morton Allport's skeletons after they travelled to Europe?

James Scott's students kept Mary McLauchlan's human remains for three days after her death. Then what was left of her was buried in an unmarked place. No religious ceremony was performed, although Reverend Bedford had made a ceremony over the dissected remains of other murderers, and in England there was an assumption that the body of a penitent murderer would be buried with some ritual. The Reverends Bedford and McArthur had abandoned her. Their neglect made something of a mockery of the hard work Bedford had boasted of putting into reclaiming her soul.

It is probable that Mary McLauchlan's family in Scotland never learned what became of her, for soon after she had been imprisoned in Glasgow's Tolbooth, her husband had suddenly disappeared from the family's dwelling. Her children would have been placed in the local workhouse, to survive as best they could in that pitiless system.

As for William Lanney, the whereabouts of his skull will probably always remain a mystery. William Crowther seems never to have sent the skull to Europe. He had no wish to seem a fool in British scientific eyes, and that would have been the effect of making such a donation once Morton Allport and James Agnew began to disseminate the rumour that Lanney was not of fully Tasmanian Aboriginal descent. Most probably the skull was still in the Crowther family home at the time of Crowther's death, and in all the unpleasant family fall-out that followed, it was disposed of like all his other possessions, as his disenchanted widow set about the business of removing herself from the colony and returning to Britain.

Of the four people whose skeletons Morton Allport shipped to Britain, only one now remains. The two he sent to London's Royal College of Surgeons were destroyed when German bombs hit the College in 1941, as was his gift to Joseph Barnard Davis, which the College had acquired the year before the skull collector's death. The man whose bones Allport sent to the Anthropological Society finally came to lie in the British Museum of Natural History, and this is the only Tasmanian skeleton known to remain in Britain.

The British Museum is among the most recalcitrant in repatriating Aboriginal skeletal material, in the interests of 'research' that seems endlessly deferred. Joseph Barnard Davis's nineteenth-century assessment of the usefulness of that museum's collection reveals how lightly the argument that such remains must be kept for 'science' should perhaps be taken. In 1867 he said the museum had cellars that were full of these precious scientific objects, which had been so 'ingeniously cata-logued' that it was 'difficult to ascertain the number even of the specimens!'.[1] Others agreed. The museum's collection was said to be 'composed of skulls filthy with dust, and in a dark cellar'.[2] So much for the argument that some kind of dis-embodied research-based science required these bones.

As for Thomas Ross, nobody knows the whereabouts of his decapitated body, of which there is no trace in cemetery records. However, there is an intriguing ref-erence to his adventurous skull. After its hasty reburial in St David's Cemetery on the morning it was discovered in 1869, somebody returned and dug it up, then quietly reinterred it in Trinity Anglican Cemetery.[3] If Ross's Presbyterian beliefs were strong, he would not have liked that final resting place. But during the twen-tieth century, when this cemetery was closed, it was turned into a playground for a local primary school. The old schoolmaster thus came to lie beneath the skipping feet of children. There is some symmetry in that.

DEALINGS WITH THE DEAD

Exploring the episodes in this book has revealed that far more than the making of medical knowledge took place in nineteenth-century dissecting rooms, and in these moments of excess, authoritative identities were created and maintained. Giovanni Aldini conducted scientific experiments that held the possibility of resuscitating the dead, and yet could not resist impressing his audience by performing a puppeteer's tricks on George Foster's body. Robert Knox preserved Mary Paterson in whisky for a later instructive dissection, and also on paper, to capture an image of woman's perfect form. He allowed men who had been intimate with her in life to undress and pose her erotically in death.

James Scott anatomised murderers before an invited audience, in a style that was part of an ongoing quest to assert his authority over less privileged medical men and a colonial governor. And William Crowther, keenly feeling his distance from the British heart of things, sought personal recognition by cutting out William Lanney's skull with as much of an eye to evening up local scores as to impressing metropolitan science.

The resulting scandal revealed much about such men's sense of entitlement to other people's bodies in the dissecting room. As for George Stokell, he took other opportunities in reducing Lanney to a shell and souveniring sufficient of his skin to make a soft purse for a friend. That activity counted as much in the scheme of things as did his 'scientific' dissection in the charnel house he made of a back room in Hobart Town's hospital.

These men created themselves on the human remains of the poor, of Aboriginal people, and of people convicted of murder. They forged and contested their own identities in these intensely personal dealings with the dead.

RECENT SCANDALS

You may tell me that things have changed. But medicine's past suffuses its present. The most recent spate of scandals has revealed that privileged access to the dead is continually abused. Legislation is one thing, and daily practice in hospitals, universities and morgues can be another. Today, as in the nineteenth century, it is only when word of abusive practices leaks out of such places that scandals ensue and calls are made for law reform. It is as if such behaviour is being discovered for the first time.

Recent revelations began in a mundane way. In 1999 a medical scientist giving evidence to the Bristol Royal Infirmary Inquiry in England made reference to a large collection of children's hearts stored at the Royal Liverpool Children's National Health Service Trust Hospital (Alder Hey). Parents were horrified to learn that doctors who had performed post-mortem examinations on their children had systematically removed organs from their bodies without consent. When the bodies were subsequently returned to relatives for burial, they may have been little more than empty shells.

During the enquiry into practices at Alder Hey, it became apparent that such unlawful and unethical practices were taking place at medical institutions throughout Britain. Despite the existence of the *Anatomy Act, 1984* and the *Human Tissue Act, 1961*, bodies were routinely being harvested without consent during both hospital and coronial post-mortem examinations. Enquiry reports reveal how little notice some medical personnel have needed to take of the legislation, regulations and codes of practice that set out the terms upon which the living may engage with the dead.

In providing evidence to these enquiries, many medical scientists have claimed to be ignorant of the law. Some seemed puzzled by the tumult, explaining they were only acting in ways that were standard practice in hospitals, universities and other research facilities. Others attempted to lay the blame at the feet of a few maverick researchers, such as Alder Hey pathologist Dr Dick van Velzen, who was an avid collector. He removed every organ in every case upon which he worked and retained them, though very little actual research was carried out on these specimens.

But tracing the network of relationships through which human remains passed from hospital beds and morgues to boxes, jars and tissue slides reveals that the people who had access to these bodies held a shared belief that the dead should be of use to the living. A 'culture of expectation' facilitated such bodily harvesting, which took place in routine ways. Somebody on a ward at a women's hospital would telephone an interested researcher to let them know that a foetus was 'available for collection' after a miscarriage or abortion. During a coronial post-mortem examination of a child's body, a pathologist would allow the heart to be transferred to a research collection. It was part of generally accepted practice. In one doctor's words, 'it was expected that one would retain the heart and lungs', and so it 'would have taken a very brave person not to have sent those specimens through'. There was also an intention to deceive, as when one of van Velzen's assistants telephoned a medical scientist to ask how best she could disguise from parents the fact that their children's eyes had been removed.

The Alder Hey enquiry revealed that such collecting long predated van Velzen's arrival in Liverpool in 1988. Eyes, hearts, lungs, heads and even the whole bodies of children and foetuses were stored at the Institute of Child Health. Some had been there for so long that they seem to have been forgotten. Parents who had been asked if they would donate their children's bodies to 'research' had been assured that the bodies would subsequently be buried, but those burials did not take place.

The remains rediscovered in 1999 included, among numerous foetuses in jars, one labelled 'Inflated monster. Humpty Dumpty', and another with the notice: 'Neck deeply lacerated. Pull it to pieces some time and reject'. Many were not labelled at all. A child's body filled one jar, and the separated head another. The Alder Hey report noted, 'The simple fact is that the retention of organs was commonplace over the country. Doctors had been brought up to expect it'.[4]

In April 2000 another scandal arose. Mrs Elaine Isaacs learned, to her horror, that her husband's brain had been removed and sent elsewhere during a coronial post-mortem examination carried out to establish the cause of his death sixteen years earlier in Manchester. Unlike hospital post-mortems, no consent is required to perform a coronial autopsy; however, during that procedure no parts may legally be removed except for the purpose of establishing the cause of death. Mrs Isaacs belatedly learned that her husband's brain had been sent to researchers at Manchester University for their use. Again, the subsequent investigation revealed the facilitating links between coronial staff and university researchers through which such material travelled.[5]

Ruth Richardson has pointed to a 'fearful symmetry' between medicine's past and its present. The demand for human remains far outstrips supply. The bodies of most value to science are those of people who have died very recently. Grieving relatives can feel pressured to consent to their use and, when consent is not given (or has not even been requested), bodies are used unlawfully, as in the past. They are treated as commodities, for a body that has been broken into its constituent parts has a very high market value, and although profiting from its sale is unlawful that nevertheless takes place. University staff have been caught selling donated human remains to enterprises that deal in such material for commercial gain. This is a confluence of factors of the kind that Britain's burkers took advantage of all those years ago.[6]

In the midst of such scandals, Gunther von Hagens arrived in London with his collection of human remains and set up business in the Atlantis Gallery in Brick Lane in March 2002. His exhibition comprised some twenty-five plastinated bodies and an additional 175 body parts. It was very successful, as it had been in other countries in which it had been displayed since 1996. Both anatomist and showman, von Hagens informed reporters, 'I don't mind if you're sensationalist in your article. More people will come if you are.'[7] He took deliberate advantage of the opportunities for publicity that were provided by Britain's body parts scandals. Outraging many, he invited to the exhibition the parents of the children whose organs had been retained at Alder Hey. And as if that were not sufficiently provocative, on hearing that the government might be drafting new laws to ban Body Worlds, he staged that public 'autopsy' on Peter Meiss.

The government subsequently published a code of practice regarding the import and export of human body parts. Its wording revealed that human remains ('fresh, frozen, plastinated, dried, embalmed or preserved in some way') were travelling around the world, sometimes through the postal service, in an unregulated trade involving buyers and suppliers abroad. The code even placed the import and export of human remains as being subject to 'world trade rules and EC law relating to the free movement of goods and services'.[8]

These current events illustrate that we will not fully understand how abusive practices with the dead take place if we limit ourselves to law reform, for the abuses occur within long-established, taken-for-granted practices in hospitals, morgues, universities and other research facilities, in which some people's bodies are turned into other people's possessions. This long history has been lost in the enquiries into current abuses, in which the only mention of this dark aspect of medicine's past is some brief reference to grave-robbing and the murders carried out by Burke and Hare. Yet as this book has revealed, the only way to understand these abuses is to take into account the culture that operates in dissecting rooms and anatomy theatres.

During the nineteenth century, this culture encouraged some practitioners to believe they were entitled to make whatever uses of the dead they wished. We need to fully understand this cultural phenomenon in order to be able to participate in ongoing debates about the use and abuse of human remains, and understand what will deter unlawful and unethical dealings with the dead.

ENDNOTES

ABBREVIATIONS

AL — Allport Library and Museum of Fine Arts, Hobart

AOT — Archives Office of Tasmania, Hobart

CL — Crowther Library, State Library of Tasmania, Hobart

DL — Dixson Library, State Library of New South Wales, Sydney

GHA — Guy's Hospital Archives, London

ML — Mitchell Library, State Library of New South Wales, Sydney

NLA — National Library of Australia, Canberra

RAI — Royal Anthropological Institute of Great Britain and Ireland, London

RCS — Royal College of Surgeons of England

RST — Records of the Royal Society of Tasmania, University of Tasmania Library, Hobart

SLNSW — State Library of New South Wales, Sydney

SLT — State Library of Tasmania, Hobart

SRO — Scottish Records Office, Edinburgh

INTRODUCTION: PERFORMING ANATOMY

1 'An Autopsy Show', *New York Times*, 23 November 2002.
2 Kass, cited in Jones, *Speaking for the Dead*, p. 4.
3 Gibbons, 'From Body Worlds to a Public Autopsy'.
4 Professor Deakin, cited in Metters, *The Isaacs Report*, p. 146.
5 Smedley, 'Habeas Corpus'.
6 *Goodbye Body Worlds Magazine*, 2003, pp. 6, 7.
7 Greteman, 'An Anatomy of Our Selves'.
8 Batty, 'Science it Wasn't'.
9 Baker, cited in Cowell, 'Public Autopsy Leaving Londoners Squirming and Gaping'.
10 Cited in Haslam, *From Hogarth to Rowlandson*, p. 256.
11 Roth, 'London's Cadaver: a Reality Show Low?'.
12 Reifler, '"I actually don't mind the bone saw"', p. 189; Capozzoli, 'A Rip into the Flesh, a Tear into the Soul', pp. 320, 318.
13 Reifler, '"I actually don't mind the bone saw"', pp. 188–9.
14 Cole, *Things for the Surgeon*.
15 Warner, *Against the Spirit of System*.

I. COMPANIONS WITH THE DEAD

1 Cited in Warner, *Against the Spirit of System*, p. 225.
2 South, *The Dissectors' Manual*, p. xv.
3 GHA, Waddington, 'Lectures on Anatomy', Vol. III, p. 2.
4 Cited in Warner, *Against the Spirit of System*, p. 236.
5 Cited in R. C. Brock, *The Life and Work of Sir Astley Cooper*, p. 15.
6 Aldini, *An Account of the Late Improvements in Galvanism*, pp. 67–8, 189–90, 191.
7 Ibid., p. 194.
8 *The Times*, 24 January 1803.
9 Aldini, *An Account of the Late Improvements in Galvanism*, p. 201.
10 Ferrari, 'Public Anatomy Lessons', p. 70.
11 RCS, Minutes, Court of Assistants 1811 to 1820, p. 86.
12 RCS, 'Record of Bodies', 18 May 1812.
13 RCS, 'Clift's Heads of Murderers', 18 May 1812.
14 RCS, 'Record of Bodies', 18 May 1812.
15 RCS, Minutes, Court of Assistants, 1811 to 1820, p. 375.
16 Bell, *A System of Dissections*, p. 101.
17 GHA, Waddington, 'Lectures on Anatomy', Vol. II, p. 253.
18 GHA, Waddington, 'Lectures on Midwifery', p. 41.
19 *Evening Mail*, 14 to 16 April 1828.
20 Green, *The Dissector's Manual*, p. 113.
21 Farr, *Elements of Medical Jurisprudence*, p. 54.
22 Ibid.

23 South, *The Dissector's Manual*, p. 279.

24 Bell, *System of Dissections*, p. xi.

25 RCS, unsourced newspaper clipping, 'Clift's Heads of Murderers'.

26 Ibid.

27 Ibid.

28 RCS, undated note on card, 'Clift's Heads of Murderers'.

29 Bell, *System of Dissections*, vol. pp. 14–15.

30 Cited in Richardson, *Death, Dissection*, pp. 30–1.

31 Cited in Rae, *Knox*, p. 64.

32 Ibid., p. 93.

33 19 April 1828, Vol. I, No. 20, p. 601.

34 Smith, 'A Lecture', p. 71.

35 31 May 1828, Vol. I, No. 26, pp. 792–3.

36 Newman, *The Evolution of Medical Education*, p. 106.

37 Fissell, *Patients, Power*, p. 163.

38 Lord Cockburn, cited by Rae, *Knox*, p. 56.

39 Cited in Roughhead, *Burke and Hare*, p. 396.

40 Newman, *Evolution*, p. 41.

41 Cited in Hull, Vicary Lecture, p. 2.

42 Cited in Newman, *Evolution*, p. 43.

43 Ibid., p. 42.

44 Bashford, *Purity and Pollution*, p. xvi.

45 All cited in Newman, *Evolution*, pp. 41, 42, 44–5.

46 Cited in Bashford, *Purity and Pollution*, p. 113.

47 Cited in Kemp, *Dr William Hunter at the Royal Academy of Arts*, p. 38.

48 Knox, *Manual*, p. 10.

49 Knox, *Races of Man*, pp. 34, 35.

50 Lonsdale, *A Sketch of the Life*, p. 101.

51 Knox, *Manual*, pp. 22, 111, 138, 115.

52 Jordanova, 'Natural Facts'.

53 Knox, *Races of Man*, p. 278.

54 Richardson, *Death, Dissection*, p. 157.

55 Ibid., p. 128.

2. DISSECTING MARY MCLAUCHLAN

1 Elkins, *The Object Stares Back*.

2 Davis, *The Tasmanian Gallows*, p. 42.

3 Cited in Rae, *Knox*, p. 88.

4 Macquarie was the Governor of New South Wales. Cited in Rimmer, *Portrait of a Hospital*, p. 4.

5 AOT, Colonial Secretary's Office, Correspondence Files, Scott to Arthur, 23 October 1832.

6 AOT, Colonial Secretary's Office, Correspondence Files, Scott to Arthur, 19 June 1834.

7 On Bock's post-mortem portraiture, see Paffen, 'The Bushrangers of Van Diemen's Land', pp. 29–38.

8 Burney cited in Porter, 'Hospitals and Surgery', p. 219.

9   Dening, *The Death of William Gooch*, pp. 13–16; Clendinnen, *Reading the Holocaust*, p. 21.

10 SRO, Presentment against Mary McLachlan.

11 AOT, Convict Department, Conduct Registers of Female Convicts.

12 AOT, Convict Department, Surgeon's Report, *Harmony*.

13 Oxley, *Convict Maids*, p. 26.

14 Foucault, *Discipline and Punish*, p. 189.

15 These quotes are from Tardif, *Notorious Strumpets*, Convict Women Nos 1504, 1506, 1574, 1510, 1551, 1565, 1561, 1575 respectively.

16 CL, uncatalogued material, 'Copies of Correspondence between Secretary of State for Colonial Department and Governors of the Australian Provinces on the Subject of Secondary Punishment'.

17 Daniels, *Convict Women*; Damousi, *Depraved and Disorderly*. On the historiography of representations of convict women, see Lake, 'Convict Women as Objects of Male Vision'.

18 *Journal of Mrs. Fenton*, pp. 354, 374; West, *History of Tasmania*, p. 509; Payne, 'A Statistical Study,' p. 56.

19 *Hobart Town Gazette*, 3 October 1829.

20 *Hobart Town Courier*, 24 April 1830.

21 AOT, Colonial Secretary's Office, Report of Board to Investigate Colonial Medical Establishment, 21 December 1831, Evidence from James Scott, pp. 277–8.

22 Hunter, *On the Uncertainty of the Signs of Murder*, p. 17.

23 *Hobart Town Courier*, 18 September 1828.

24 Farr, *Elements*, pp. 59, 62.

25 Crawford, 'A Scientific Profession', pp. 211, 213.

26 Chapman, *The Diaries and Letters*, pp. 339, 348.

27 AOT, Supreme Court, Register of Prisoners Tried in Criminal Proceedings, Sixth Session.

28 *Historical Records of Australia* [hereafter *HRA*], pp. 437, 434–8.

29 *Journal of Mrs. Fenton*, p. 351.

30 AOT, Executive Council, 16 April 1830.

31 Ibid., 17 April 1830.

32 Chapman, *The Diaries and Letters*, p. 22.

33 Chapman, *The Diaries and Letters*, p. 430.

34 AOT, Executive Council, Minutes of Meetings, 17 April 1830.

35 *Hobart Town Courier*, 24 April 1830.

3. INTERLUDE

1   Wolfe, *Settler Colonialism*, p. 27.

2   Arthur to Dixon, cited in Levy, *Governor George Arthur*, p. 104.

3   *HRA*, pp. 85–6.

4   Wolfe, *Settler Colonialism*, p. 2.

5   Arthur's letter and the report, from which the following quotations are taken, can be found in Tasmanian Historical Research Association, *Van Diemen's Land*.

6   Cited in Levy, *Governor George Arthur*, pp. 106, 110.

7   Ibid., p. 112.

8   NLA, Emmett, 'Reminiscences'.

4. THE BONE COLLECTORS

1 Anon., 'Barnard Davis', p. 387.

2 *Dictionary of National Biography* [hereafter *DNB*], pp. 618–19.

3 Davis, *Thesaurus*, p. v.

4 Knox, *Great Artists*, p. 141.

5 Beddoe, *Memories*, p. 241.

6 These quotes are all from RAI, 'Notae', 20 June 1861.

7 RAI, Notebook No. 6, p. 82.

8 RAI, uncatalogued material, 'Information Respecting the Inhabitants'.

9 Gould, *Mismeasure*, p. 83.

10 RAI, 'Various Notes', Durlyneux to Davis, 19 April 1859.

11 Beddoe, *Memories*, p. 205.

12 Davis, 'Hints for Collecting', p. 386.

13 Stocking, *Race, Culture and Evolution*, p. 59.

14 Broca, *Hybridity in the Genus Homo*, pp. 47, 49, ix.

15 Cited in Turnbull, 'Enlightenment Anthropology', p. 214.

16 Davis, *Thesaurus*, p. vi.

17 Davis, *On the Osteology*, p. 18.

18 Dubow, *Scientific Racism*, p. 23.

19 Anon., 'Anthropological News', p. 216.

20 RAI, 'Notae', 28 June and 1 July 1859.

21 Taylor, *The Diary of a Medical Student*, pp. 14–16.

22 Flower, 'Anthropology', p. 238.

23 Wyman, 'Observations', pp. 330–5.

24 RAI, Notebook No. 6, pp. 71–2.

25 Cited in Gould, *Mismeasure*, p. 105.

26 In his controversial history, Keith Windschuttle questions the veracity of this report, which was contained in colonial chaplain Robert Knopwood's diary (Windschuttle, *Fabrication*, p. 24). Nobody imbued in contemporary accounts of European bone collecting would do so. Just look at the names of all those donors in Brian Plomley's exhaustive list of the Tasmanian skeletal material that came to reside in European collections during the nineteenth century (Plomley, 'A List'). Medical men were pre-eminent bone collectors.

27 Cited in Rae-Ellis, *Trucanini*, p. 133.

28 Mauss, *The Gift*.

29 Cited by Home, *Australian Science*, p. x.

30 RCS, Museum Letters, Vol. II, Crowther to Flower, 8 October 1869; ML, Parkes Correspondence, Crowther to Henry Parkes, 17 May 1871.

31 RCS, Museum Letters, Vol. I.

32 RCS, Museum Letters, Vol. I, 23 May 1864.

33 Crowther forwarded the calf, which he estimated to be ten months old, to Flower in June 1866 (RCS, Museum Letters, Vol. I, Crowther to Flower, 24 June 1866). Flower later lamented the 'ruthless destruction' of whales at their breeding places, where the cows came close to shore to give birth and, when attacked, refused to leave their calves ('Whales and Whale Fisheries', in his *Essays*, p. 206).

34 RCS, Museum Letters, Vol. I, Flower to Crowther, 3 March 1864.

35 He wrote that in 1867 the Colonial Treasurer had given an order 'that the next [Aboriginal person] that should be taken ill is to be forwarded to the General Hospital where I need hardly say she will receive every attention at my hands, particularly post mortem'. This sentence sent shivers down my spine when I read it. However, I do not think Crowther meant to hasten the next death in order to undertake his post-mortem work. In the following paragraph he speaks of his unwillingness to disturb Aboriginal graveyards, in the knowledge that 'these poor people … have great repugnance at any disturbance of the remains of their friends'. The situation called, he said, for the passing of time and 'some little tact' (RCS, Museum Letters, Vol. I, Crowther to Flower, 23 February 1867).

36 RCS, Museum Letters, Vol. I, Flower to Crowther, 23 May 1864.

37 RCS, Museum Letters, Vol. I, Crowther to Flower, 23 February 1867.

38 Trollope, *Australia and New Zealand*, Vol. 2, pp. 162–3.

39 AL, 'Journal of Morton Allport'.

40 AL, Letter Book 1871–1874, Allport to institute, 23 October 1873.

41 AL, ibid., Allport to Davis, 23 January 1873.

42 AL, ibid., Allport to Davis, 8 August 1873.

43 *Launceston Examiner*, 11 May 1876.

44 Cited in McGregor, *Imagined Destinies*, pp. 29, 28.

45 Tylor believed the Tasmanians had existed for aeons in their present state, 'left behind in industrial development' even by the ancient tribes of Europe (Tylor, 'On the *Tasmanians* as Representatives of *Paolaeolithic Man*', pp. 148–9; Tylor cited in Stocking, *Race, Culture and Evolution*, p. 76).

46 Cited by Rae-Ellis, *Trucanini*, p. 27.

47 Flower, 'Anthropology', p. 242.

48 Flower, *Aborigines of Tasmania*, p. 47.

49 Flower, *Catalogue I*, p. viii.

50 In Lubbock, *Origin of Civilisation*, p. 345.

51 Cited in Cornish, *Sir William Henry Flower*, p. 107.

52 Flower, 'On the Native Races', p. 48.

53 Ibid., p. 6.

54 Ibid., pp. 48–9.

5. DEATH AND DISSECTION, 1869

1 In chronological order, these phrases are taken from Roth, *Aborigines of Tasmania*; Rae-Ellis, *Trucanini*; Pybus, *Community of Thieves*; Robson, *A History of Tasmania: Volume II*; Cove, *What the Bones Say*; and Petrow, 'The Last Man'.

2 Ryan, *Aboriginal Tasmanians*.

3 CL, uncatalogued material, Agnew, 'Views rê the Management of the General Hospital Hobart 1859'.

4 *Mercury*, 29 October 1864.

5 Ibid.

6 RCS, Museum Letters, Vol. II, Crowther to Flower, 14 June 1873, 11 July 1873.

7 Cited in Petrow, 'The Last Man', p. 24.

8   *Mercury*, 5 March 1869.

9   RCS, Museum Letters, Vol. II, Crowther to Flower, 27 March 1869.

10 Ibid., Crowther to Flower, 17 June 1869.

11 'Et Tu! Brute!!', reprinted in *Mercury*, 12 March 1869.

12 *Mercury*, 13 March 1869.

13 *Cornwall Chronicle*, 17 March 1869.

14 *Mercury*, 13 March 1869.

15 Ibid., 15 March 1869.

16 *Tasmanian Times*, 16 March 1869.

17 *Mercury*, 12 March 1869.

18 *Tasmanian Times*, 18 March 1869.

19 Letter to the Editor, *Cornwall Chronicle*, 12 March 1869.

20 Petrow, 'The Last Man'.

21 Letter to the Editor, *Cornwall Chronicle*, 12 March 1869.

22 *Cornwall Chronicle*, 10 April 1869.

23 Reprinted in the *Tasmanian Times*, 13 March 1869.

24 Letter to the Editor, *Cornwall Chronicle*, 31 March 1869.

25 *Tasmanian Times*, 10 March 1869.

26 Ibid., 18 March 1869.

27 CL, 'Historical Records Volume 1'.

28 Flower and Murie, 'Account of the Dissection'.

29 22 March 1869.

30 RCS, Museum Letters, Vol. II, Crowther to Flower, 22 April 1869, p. 18.

31 Letter to the Editor, *Cornwall Chronicle*, 12 March 1869.

32 Letter to the Editor, 19 March 1869.

33 *Cornwall Chronicle*, 28 August 1869.

34 Cited in *Mercury*, 13 March 1869.

35 Knox, letter to the *Caledonian Mercury*, in Roughhead, *Burke and Hare*, pp. 275–7.

36 *Mercury*, 9 March 1869.

37 *Tasmanian Times*, 13 March 1869.

38 Letter dated 8 March, published in *Mercury*, 9 March 1869.

39 *Tasmanian Times*, 15 March 1869, 17 March 1869.

40 Ibid., 6 March 1869.

41 *Mercury*, 27 March 1869.

42 Letter to the Editor, *Mercury*, 10 April 1869.

43 Letter to the Editor, *Cornwall Chronicle*, 12 March 1869.

44 RCS, Museum Letters, Vol. II, 6 November 1869.

45 Ibid., 29 January 1869.

46 Ibid., 27 March 1869.

47 AL, Letter Book August 1864 to June 1868, Allport to Mrs Butler, 22 March 1866.

48 Ibid., Allport to unnamed woman, 25 September 1867.

49 Ibid., Allport to Henslowe, 22 April 1865.

50 AL, Letter Book 1868 to 1872, Allport to Flower, 29 December 1871.

51 Ibid., 11 July 1872.

52 AL, Letter Book November 1872 to December 1874, Allport to Davis, 8 August 1873.

53 Ibid.

54 RCS, Museum Letters, Vol. II, Crowther to Flower, 19 April 1873.

55 Ibid., 22 April 1869.

56 Agnew, 'The Last of the Tasmanians'.

57 Parliamentary Papers, Report of Select Committee, September 1875, p. 1.

58 Related in Petrow, 'The Last Man', pp. 39–40.

6. THE USE AND ABUSE OF HUMAN REMAINS

1 Davis, *Thesaurus*, p. xii.

2 Anon., 'Barnard Davis', pp. 387, 338.

3 Personal communication, Margaret Glover, AOT.

4 *Royal Liverpool Children's Inquiry*, pp. 90–91, 114, 284.

5 Metters, *The Isaacs Report*.

6 Richardson, *Death, Dissection*, 2001, pp. 410, 413–14.

7 Jeffries, 'The Naked and the Dead'.

8 Department of Health, *The Import and Export of Human Body Parts*, p. 6.

BIBLIOGRAPHY

ARCHIVAL MATERIAL

In Britain

Guy's Hospital Archives, London

Waddington, Joshua, 'Lectures on Anatomy; and the Principle Operations of Surgery. Delivered at the Theatre, St Thomas's Hospital; between the 1st of January and the 1st of June, 1816 by Astley Cooper Esq.', 3 volumes.

——, 'Lectures on Midwifery and Diseases of Woman, and Children: Delivered at the Theatre, Guys Hospital between the 1st of November 1816 and the 1st of March 1817 by John Haighton, M.D'.

Public Records Office, London

Colonial Office, Despatches January 1830 to May 1830, CO280/24.

Royal Anthropological Institute of Great Britain and Ireland, London

Joseph Barnard Davis Papers

Notebook No. 6, September 1859 to 1 October 1860, MS140:6.

'Notae Ethnographicae, 1859–1865', MS147.

'Various Notes, Observations, References etc. Medical & Archaeological, Partly under Alphabetical Heads', Halifax, c. 1821 to 1868, MS145.

Uncatalogued material: Correspondence 1854, 1865; 'Extract from a Letter to the Archbishop of Dublin from Thomas Arnold, D. D., Rugby, March 22 1835'; 'Information Respecting the Inhabitants of the British Isles', leaflet printed for the author, 1 January 1861; 'Notes on the Preparation of Crania in Hot Climates, and Chiefly Applicable to India'.

RAI House Archives
Council Minutes of the Anthropological Institute 1871–82, A10:1.

Royal College of Surgeons of England
'Clift's Heads of Murderers', 1807–1832, Box 67.b.13.
Copies of Two Letterbooks 1800–1850, typescript, E2b and E2c.
Minutes, Court of Assistants, Vol. III, 1811 to 1820.
Museum Letters, Vol. I, 1862–68; Vol. II, 1868–1873; Vol. III, 1874–78.
'Record of the Bodies of Murderers Delivered to the College for Dissection'.
William Clift's Diaries.

Scottish Records Office, Edinburgh
First Declaration of Mary McLachlan, 9 November 1827, JC26/520.
Precognition Relative to a Charge of Housebreaking and Theft or Reset of Theft by M. McLachlane, 9 November 1827, AD/14/28/203.
Presentment against Mary McLachlan, 1828, AD14/28/161.
Third Declaration of Mary McLachlane, 1 December 1827, JC26/520.

In Australia

Allport Library and Museum of Fine Arts, Hobart
Allport, Morton, 'Specimens forwarded to Royal Museum of Natural History, Brussells, Belgium, December 18th 1872', Box 1.
——, Letter Books, August 1864 to June 1868; 1868 to 1872; November 1872 to December 1874.
——, 'Journal of Morton Allport during his stay at Paris & other places on the continent—1854, copied out by Eva Mary Allport', Box 1.

Archives Office of Tasmania, Hobart

Colonial Secretary's Department
Correspondence files, du Cane period, CSD7/23/127.

Colonial Secretary's Office
Correspondence Files, Arthur period, CSO1/514/112277, CSO1/616/14050–14081.
Report of Board to Investigate Colonial Medical Establishment, 21 December, 1831, CSO1/582/13172.
Report and Evidence to Commission on Female Convict Discipline, 1841–43, CSO22/50.
Report on the Present State of the Nurseries and Hospital at the House of Correction for Females, 1834, CSO1/902/19161.
Shipping List, *Harmony*, CSO1/3/368/8375.

Convict Department
Appropriation List of the Ship *Harmony*, MM33/5.
Assignment List of the Ship *Harmony*, MM33/3.
Conduct Registers of Female Convicts, CON40/5.
Description Lists of Female Convicts, CON18/24.
Surgeon's Report, *Harmony*, Admin 101/32.

Executive Council
Minutes of Meetings, Arthur period, EC4/1, 2.

Lieutenant Governor's Office
Correspondence Files, Arthur Period, GO33/7.

Supreme Court
Criminal Minute Book for Van Diemen's Land and its Dependencies, 24 January 1829 to 23 October
    1829, SC32/1.
Register of Persons Tried in Criminal Proceedings, Fourth Session, Oatlands, 1842, SC41/5.
Register of Prisoners Tried in Criminal Proceedings, Sixth Session, SC41/3.

Miscellaneous
Pioneer Index, Register of Burials.
Purtscher, Joyce (compiler), 'Deaths at General Hospital Hobart January 1864–June 1884', unpublished.

Crowther Library, State Library of Tasmania, Hobart

Scrapbooks
'Historical Records Volume 1, compiled by A. W. Hume Esq., late editor of Tasmanian News and
    Critic', scrapbook of unsourced newspaper clippings purchased by Sir William Crowther, Box
    PQ362.2 365.
Anonymous, 'The Last Man of a Race', Scrapbook, Box PQ362.2 365.

Uncatalogued Material
'A scrapbook compiled by Sir William Crowther, 1950', Commissioners for Tasmania to Mrs Crowther,
    16 December 1861.
'Copies of Correspondence between Secretary of State for Colonial Department and Governors of the
    Australian Provinces on the subject of secondary punishment, House of Commons, 4 March
    1834'.
Sir J. Agnew, formerly assistant colonial surgeon, papers relating to his activities mostly medical, Van
    Diemen's Land, 'Views rê the Management of the General Hospital Hobart 1859', ms.

National Library of Australia, Canberra
Cotton, Reverend, 'Original Diary or Visiting Book of the Reverend H. S. Cotton, Chaplain of Newgate,
    from 1823 to 1836, detailing day by day his Visits to the Prison, Conversations with Prisoners,
    their Crimes, Last Hours of the Condemned, their Executions, Confessions, & etc', MS14.
Emmett, Henry James, 'Reminiscences of the Black War in Tasmania', 1873, MS3311.

Royal Society of Tasmania, Hobart

Letter Books, January 1874 to 25 September 1885, RSA/B/2.

Minutes of Council and Monthly Society Meetings from 1845, 1863–78, Vol. 4, RSA A1–9.

State Library of New South Wales, Sydney

Dixson Library, Tasmanian Material

Coroners' Inquests, Arthur Period, ADD554, 601.

Tasmania, Police Courts, Chief Police Magistrate, Hobart, Records 9 July 1829 to 31 March 1830, Vol. 1, 64/434.

Mitchell Library

Tasmania Police, Supreme Court, Information, depositions, & etc., 1829–48, Tasmania Papers 236.

Parkes Correspondence, Vol. 46.

PUBLISHED WORKS

Agnew, J. W., 'The Last of the Tasmanians', *Proceedings of Australasian Association for the Advancement of Science*, Sydney, 1888.

Aldini, John (Giovanni), *An Account of the Late Improvements in Galvanism, with a series of various and interesting experiments performed before the Commissioners of the French National Institute, and repeated lately in the Anatomical Theatres in London (to which is added an Appendix, containing the author's experiments on the body of a malefactor executed at Newgate &c., &c.)*, Cathell & Martin, London, 1803.

Aldini, Giovanni, *Essai théorique et expérimental sur le galvanisme*, Fournier, Paris, 1804.

'An Autopsy Show', *New York Times*, 23 November 2002.

Anon, 'Anthropological News', *Anthropological Review*, Vol. V, No. xvi, 1867.

——, 'Barnard Davis on Cranioscopy', *Anthropological Review*, Vol. VI, No. xxiii, 1868.

——, *Companion to Post-Mortem Examinations*, John Rose & William White, London, 1831.

Ariès, Phillipe, *The Hour of our Death*, Allen Lane, London, 1981, translated by Helen Weaver.

Baillie, Matthew, *The Morbid Anatomy of Some of the Most Important Parts of the Human Body*, W. Bulmer & Co., London, 1807, third edition.

Barzun, Jacques (ed.), *Burke and Hare the Resurrection Men: A Collection of Contemporary Documents including Broadsides, Occasional Verses, Illustrations, Polemics, and a Complete Transcript of the Testimony at the Trial*. From the Fenwick Beekman Collection at the New York Academy of Medicine, Scarecrow Press, Metuchen, NJ, 1974.

Bashford, Alison, *Purity and Pollution: Gender, Embodiment and Victorian Medicine*, Macmillan Press, Hampshire, 1998.

Bateson, Charles, *The Convict Ships 1787–1868*, Brown, Son & Ferguson, Glasgow, 1959.

Batty, David, 'Science it Wasn't, but What an Anecdote for a Dinner Party', *Guardian Unlimited*, http://society.guardian.co.uk/news/story/0,7838,844928.00.html, consulted 29 March 2004.

BBC News, 'Body Doctor Cleared over Corpses', http://news.bbc.co.uk/2/hi/entertainment/3497306.stm, consulted 29 March 2004.

——, 'Russians Charged over Body Parts', 20 July 2002, http://news.bbc.co.uk/2/hi/europe/2140333.stm, consulted 29 March 2004.

Beddoe, John, *Memories of Eighty Years*, J. W. Arrowsmith, Bristol, 1910.

Bell, Charles, *A System of Dissections Explaining the Anatomy of the Human Body, the Manner of Displaying the Parts, and their Varieties in Disease*, Volumes 1 and 2, Mundell & Son, Edinburgh, 1799, second edition.

Bloch, Marc, *The Historian's Craft*, Manchester University Press, Manchester, 1954, translated by Peter Putnam.

Bock, Thomas, *Convict Engraver, Society Portraitist*, Queen Victoria Museum and Art Gallery & Australian National Gallery, Launceston and Canberra, 1991.

Body Worlds website, 19 November, 2002, http://www.bodyworlds.com/en/pages/specials_london_autopsie.asp, consulted 29 March 2004.

Bonwick, James, *The Last of the Tasmanians or, the Black War of Van Diemen's Land*, Sampson Low, Son & Marston, London, 1870.

Brand, I. and Staniforth, M., 'Care and Control: Female Convict Transportation Voyages to Van Diemen's Land, 1818–1853', *The Great Circle*, No. 16, 1994.

Briggs, Asa, *The Age of Improvement 1783–1867*, Longmans, London, 1959.

Bristol Royal Infirmary Inquiry, *The Report of the Public Inquiry into Children's Heart Surgery at the Bristol Royal Infirmary 1984–1995: Learning from Bristol*, Presented to Parliament by the Secretary of State for Health by Command of Her Majesty, July 2001.

Broca, Paul, *Hybridity in the Genus Homo*, published for the Anthropological Society of London by Longman, Green, Longman & Roberts, London, 1864.

Brock, R. C., *The Life and Work of Sir Astley Cooper*, E. & S. Livingstone Ltd, Edinburgh and London, 1952.

Bryden, William, 'Tasmanian Museum and Art Gallery: Historical Note', *Papers and Proceedings of the Royal Society of Tasmania*, No. 100, 1966.

Burn, David, *A Picture of Van Diemen's Land*, Cat and Fiddle Press, Hobart, 1973, first published 1840.

Bynum, W. F., *Science and the Practice of Medicine in the Nineteenth Century*, Cambridge University Press, Cambridge, 1994.

Byrne, Paula, 'On Her Own Hands: Women and Criminal Law in New South Wales, 1810–1830', in David Philips and Susanne Davies (eds), *A Nation of Rogues? Crime, Law and Punishment in Colonial Australia*, Melbourne University Press, Melbourne, 1994.

Calder, J. E., *Some Account of the Wars, Extirpation, Habits etc. of the Native Tribes of Tasmania*, Fullers Bookshop, Hobart, 1972, originally published in 1875.

Cameron, Joy, *Prisons and Punishment in Scotland from the Middle Ages to the Present*, Canongate, Edinburgh, 1983.

Capozzoli, Maureen, 'A Rip into the Flesh, a Tear into the Soul: An Ethnography of Dissection in Georgia', in Jackson Harrington (ed.), *Bones in the Basement: Postmortem Racism in Nineteenth-Century Medical Training*, Smithsonian Institution Press, Washington DC, 1997.

Casper, Claudia, *The Reconstruction*, Quartet Books, London, 1997.

Cazort, Mimi, Kornell, Monique and Roberts, K. B. (eds), *The Ingenious Machine of Nature: Four Centuries of Art and Anatomy*, National Gallery of Canada, Ottawa, 1996.

Chapman, Peter (ed.), *The Diaries and Letters of G. T. W. B. Boyes*, Volume 1, 1820–1832, Oxford University Press, Melbourne, 1985.

Chrisafis, Angelique, 'Artist Insists his Bodies will Survive Legal Fight', *Guardian*, 12 March 2002, http://www.guardian.co.uk/print/0,3858,4372450–103690,00.html, consulted 29 March 2004.

Clark, C. M. H., *A History of Australia II: New South Wales and Van Diemen's Land 1822–1838*, Melbourne University Press, Carlton, 1968.

Clark, Julia, *This Southern Outpost: Hobart 1846–1914*, Corporation of the City of Hobart, Hobart, 1988.

Clendinnen, Inga, *Reading the Holocaust*, Text Publishing, Melbourne, 1998.

Cole, F. J., *A History of Comparative Anatomy from Aristotle to the Eighteenth Century*, McMillan & Co., London, 1944.

Cole, Hubert, *Things for the Surgeon: A History of the Resurrection Men*, Heinemann, London, 1964.

Cooper, David, *The Lesson of the Scaffold*, Allen Lane, London, 1974.

Cooter, Roger, *The Cultural Meaning of Popular Science: Phrenology and the Organization of Consent in Nineteenth-Century Britain*, Cambridge Studies in the History of Medicine, Cambridge University Press, Cambridge, 1984.

Cope, Zachary, *The Royal College of Surgeons, England: A History*, Anthony Bland Ltd., London, 1959.

Cornish, Charles, *Sir William Henry Flower K.C.B., F.R.S., LL.D., D.C.L. Late Director of the Natural History Museum, and President of the Royal Zoological Society. A Personal Memoir*, Macmillan & Co., London, 1904.

Cove, John, *What the Bones Say: Tasmanian Aborigines, Science and Domination*, Carleton University Press, Ontario, 1995.

Cowell, Alan, 'Public Autopsy Leaving Londoners Squirming and Gaping', *New York Times*, 21 November 2002, wysiwyg://20/http://www/nytimes.com/20, consulted 23 November 2002.

Craig, Clifford, *Mr Punch in Tasmania: Colonial Politics in Cartoons 1866–1879*, Blubber Head Press, Hobart, 1980.

Crawford, Catherine, 'A Scientific Profession: Medical Reform and Forensic Medicine in British Periodicals of the Early Nineteenth Century', in Roger French and Andrew Wear (eds), *British Medicine in an Age of Reform*, Routledge, London and New York, 1991.

Crowther, M. Anne and White, Brenda, *On Soul and Conscience: The Medical Expert and Crime: 150 Years of Forensic Medicine in Glasgow*, Aberdeen University Press, Aberdeen, 1988.

Crowther, W. E. L. H., 'Aspects of the Life of a Colonial Surgeon: The Hon. W. L. Crowther, F.R.C.S., C.M.Z.S., Sometime Premier of Tasmania', *Medical Journal of Australia*, Vol. II, No. 13, 1942.

——, 'The Changing Scene in Medicine: Van Diemen's Land 1803–1853', *Medical Journal of Australia*, Vol. I, 1942.

——, 'Dr E. S. P. Bedford and his Hospital and Medical School of Van Diemen's Land', *Medical Journal of Australia*, Vol. II, No. 2, 1944.

——, 'Practice and Personalities at Hobart Town, 1828–1832, as Indicated by the Day Book of James Scott, M.D., R.N., Senior Colonial Surgeon', *Medical Journal of Australia*, Vol. I, No. 12, 1954.

——, 'Some Aspects of Medical Practice in Van Diemen's Land, 1825–1839', *Medical Journal of Australia*, Vol. I, No. 17, 1935.

——, 'A Surgeon as Whaleship Owner', *Medical Journal of Australia*, Vol. I, No. 25, 1943.

Crowther, W. L., 'Urethrotomy or Lithotrity in Aged and Debilitated People', *Lancet*, 1 November 1873.

Dally, Ann, *Women Under the Knife*, Hutchinson Radius, London, 1991.

Damousi, Joy, *Depraved and Disorderly: Female Convicts, Sexuality and Gender in Colonial Australia*, Cambridge University Press, Cambridge, 1997.

Daniels, Kay, *Convict Women*, Allen & Unwin, Sydney, 1998.

Davis, Joseph Barnard, 'Hints for Collecting and Preserving the Bones of Ancient Skulls', *Gentleman's Magazine*, October, Vol. XI, 1853.

——, *On the Osteology and Peculiarities of the Tasmanian, a Race of Man Recently Become Extinct*, De Erven Loosjes, Haarlem, the Netherlands, 1874.

——, *Supplement to Thesaurus Craniorum, Catalogue of the Skulls of the Various Races of Man in the Collection of Joseph Barnard Davis, M.D., F.S.A, etc.*, printed for subscribers, London, 1875.

——, *Thesaurus Craniorum. Catalogue of the Skulls of the Various Races of Man in the Collection of Joseph Barnard Davis, M.D., F.S.A., etc.*, printed for subscribers, London, 1867.

Davis, Joseph Barnard and Thurman, John, *Crania Britannica: Delineations and Descriptions of the Skulls of the Aboriginal and Early Inhabitants of the British Islands, with Notices of their Remains*, 2 volumes, printed for subscribers, London, 1865.

Davis, Richard, *The Tasmanian Gallows: A Study of Capital Punishment*, Cat and Fiddle Press, Hobart, 1974.

Deans, Jason, 'C4 to Broadcast Autopsy', *Guardian*, 19 November 2002, http://media.guardian.co.uk/Print/0,3858,4550104,00.html, consulted 22 November 2002.

Dening, Greg, *The Death of William Gooch: A History's Anthropology*, Melbourne University Press, Carlton, 1995.

Department of Health, *Import and Export of Human Body Parts and Tissue for Non-Therapeutic Uses: A Code of Practice*, Department of Health, London, April 2003.

Desmond, Adrian, *The Politics of Evolution: Morphology, Medicine, and Reform in Radical London*, University of Chicago Press, Chicago, 1989.

Desmond, Adrian and Moore, James, *Darwin*, Penguin, London, 1992, first published 1991.

Dictionary of National Biography from the Earliest Times to 1900, Sir Leslie Steven and Sir Sidney Lee (eds), Oxford University Press, London, 1921–22.

'Dissecting the Dead for Show', *Science Now*, 25 November 2002, bric.postech.ac.ks/science/97now/02_11now/021125a.html, consulted 20 November 2003.

Dobson, Jessie, 'The Anatomising of Criminals', *Annals of the Royal College of Surgeons of England*, Vol. 9, No. 2, August 1951.

——, *William Clift*, Heinemann, London, 1954.

'Doctor Stages Public Autopsy', *CNN.com*, 20 November 2003, http://www.CNN.com/2002/WORLD/europe/11/20/uk.public.autopsy/index.html, consulted 27 November 2003.

Dubow, Saul, *Scientific Racism in Modern South Africa*, Cambridge University Press, Cambridge, 1995.

Duffield, Ian, '"Stated This Offence": High-Density Convict Micro-Narratives', in Lucy Frost and Hamish Maxwell-Stewart (eds), *Chain Letters: Narrating Convict Lives*, Melbourne University Press, Carlton, 2002.

Duffield, Ian and Bradley, James (eds), *Representing Convicts: New Perspectives on Convict Forced Labour Migration*, Leicester University Press, London, 1997.

Durey, M. J., 'Bodysnatchers and Benthamites: The Implications of the Dead Body Bill for the London Schools of Anatomy, 1820–42', *The London Journal: A Review of Metropolitan Society Past and Present*, Vol. 2, No. 2, 1976.

Edwards, Owen Dudley, *Burke and Hare*, Polygon, Edinburgh, 1984.

Elam, Gillian, *Consent to Organ and Tissue Retention at Post-mortem Examination and Disposal of Human Material*, prepared for the Health Services Directorate, Department of Health, London, 2000.

Elkins, James, *The Object Stares Back: On the Nature of Seeing*, Harcourt, Inc., San Diego, 1996.

Farr, Samuel, MD, *Elements of Medical Jurisprudence; or a Succinct and Compendious Designation of such Tokens in the Human Body as are Requisite to Determine the Judgment of a Coroner, and Courts of Law, in Cases of Divorce, Rape, Murder, &c. to which are added Directions for Preserving the Public Health*, printed for Callow, Medical Bookseller, London, 1815, third edition.

Ferrari, Giovanna, 'Public Anatomy Lessons and the Carnival: The Anatomy Theatre of Bologna', *Past and Present*, No. 117, 1987.

Fforde, C., 'English Collections of Human Remains: An Introduction', *World Archaeological Bulletin*, No. 6, 1992.

Field, Barron, 'On the Aborigines of New Holland and Van Diemen's Land, a lecture read before the Philosophical Society of Australia, 2 January 1822', in Barron Field (ed.), *Geographical Memoirs on New South Wales / by Various Hands. Containing an Account of the Surveyor General's Late Expedition to Two New Ports, the Discovery of Moreton Bay River, with the Adventures for Seven Months there of Two Shipwrecked Men, a Route from Bathurst to Liverpool Plains: Together with Other Papers on the Aborigines, the Geology, the Botany, the Timber, the Astronomy, and the Meteorology of New South Wales and Van Diemen's Land*, John Murray, London, 1825.

Fissell, Mary, *Patients, Power and the Poor in Eighteenth Century Bristol*, Cambridge University Press, Cambridge, 1991.

Flower, Sir William Henry, *The Aborigines of Tasmania an Extinct Race*, Science Lectures Delivered in Manchester 1877–79, 9th and 10th Series, John Heywood, Manchester, 1877–79.

——, 'Anthropology', from an original presentation to the British Association for the Advancement of Science at its York meeting on 1st September 1881, in William Henry Flower, *Essays on Museums and Other Subjects Connected with Natural History*, Macmillan & Co., London, 1898.

——, *Catalogue of the Specimens Illustrating the Osteology and Dentition of Vertebrated Animals, Recent and Extinct, Contained in the Museum of the Royal College of Surgeons of England. Part I: Man: Homo Sapiens, Linn.*, Royal College of Surgeons, London, 1879.

——, *Catalogue of the Specimens Illustrating the Osteology and Dentition of Vertebrated Animals, Recent and Extinct, Contained in the Museum of the Royal College of Surgeons of England. Part II: Class Mammalia, other than Man*, Royal College of Surgeons, London, 1884.

——, *Essays on Museums and Other Subjects Connected with Natural History*, Macmillan & Co., London, 1898.

——, 'On the Native Races of the Pacific Ocean', *Proceedings of the Royal Institution of Great Britain*, Vol. VIII, 1879.

——, 'On the Osteology of the Cachalot or Sperm-whale (Physeter macrocephalus)', *Zoological Society Transactions*, Vol. VI, 1869.

Flower, William and Murie, James, 'Account of the Dissection of a Bushwoman', *Journal of Anatomy and Physiology*, Vol. 1, 1867.

Foucault, Michel, *Discipline and Punish: The Birth of the Prison*, Penguin Books, London, 1977.

——, *The Birth of the Clinic: An Archaeology of Medical Perception*, Pantheon, New York, 1973.

Fox, Colonel A. Lane, 'On the Principles of Classification Adopted in the Arrangement of his Anthropological Collection, now exhibited in the Bethnal Green Museum', paper read at the Special Meeting of the Anthropological Institute of Great Britain and Ireland held at the Bethnal Green Museum, 1st July, 1874, on the occasion of the opening of the Collection to the public, *Journal of the Anthropological Institute of Great Britain and Ireland*, Vol. 4, 1874.

French, Roger and Wear, Andrew, *British Medicine in the Age of Reform*, Routledge, London and New York, 1991.

Gatrell, V. A. C., *The Hanging Tree: Execution and the English People 1770–1868*, Oxford University Press, Oxford, 1994.

Geary, Laurence M., 'The Scottish–Australian Connection 1850–1900', in Vivian Nutton and Roy Porter (eds), *The History of Medical Education in Britain*, Rodopi, Amsterdam, 1995.

Gibbons, Fiachra, 'From Body Worlds to a Public Autopsy', *Guardian*, http://www.guardian.co.uk/print/0,3858,4545279_103690,00.html, consulted 22 November 2002.

Gillen, Molly, *Assassination of the Prime Minister: The Shocking Death of Spencer Perceval*, Sidgwick & Jackson, London, 1972.

Gilman, Sander, 'Black Bodies, White Bodies: Toward an Iconography of Female Sexuality in Late Nineteenth-Century Art, Medicine, and Literature', *Critical Inquiry*, Vol. 21, No. 2, 1985.

Ginzburg, Carlo, *The Cheese and the Worms: The Cosmos of a Sixteenth-Century Miller*, Routledge and Kegan Paul, London, 1980, translated by John and Anne Tedeschi.

Goodbye Body Worlds Magazine, 2003.

Gould, Stephen Jay, *The Mismeasure of Man*, Penguin, London, 1996, revised edition.

Green, F. C. (ed.), *A Century of Responsible Government in Tasmania 1856–1956*, L. G. Shea, Government Printer, Tasmania, 1956.

Green, Joseph Henry, *The Dissector's Manual*, printed for the author, London, 1820.

Greteman, Blaine, 'An Anatomy of Our Selves', *Time Magazine*, 2 December, 2002, http://www.time.com/time/nation/article/0,18599,393606,00.html, consulted 29 March 2004.

Griffiths, Tom, *Hunters and Collectors: The Antiquarian Imagination in Australia*, Cambridge University Press, Melbourne, 1996.

Hafferty, Frederic, *Into the Valley: Death and the Socialization of Medical Students*, Yale University Press, New Haven and London, 1991.

Haslam, Fiona, *From Hogarth to Rowlandson: Medicine in Art in Eighteenth-Century Britain*, Liverpool University Press, Liverpool, 1996.

Hay, Douglas, 'Property, Authority and the Criminal Law', in D. Hay, P. Linebaugh, J. Rule, E. P. Thompson and C. Winslow (eds), *Albion's Fatal Tree: Crime and Society in Eighteenth-Century England*, Allen Lane, London, 1975.

Historical Committee of the National Trust of Australia, Tasmania, *Campbell Town Tasmania: History and Centenary of Municipal Government*, Campbell Town Municipal Council, Hobart, 1966.

Historical Records of Australia, Resumed Series, Series III, Vol. VII, 1828, Peter Chapman (ed.), Australian Government Printing Service, Canberra, 1997.

Home, R. W. (ed.), *Australian Science in the Making*, Cambridge University Press in association with the Australian Academy of Science, Cambridge, 1990.

Home, Rod and Kohlstedt, Sally (eds), *International Science and National Scientific Identity*, Kluwer Academic Publishers, Dordrecht, The Netherlands, 1991.

Hooper-Greenhill, E., *Museums and the Shaping of Knowledge*, Routledge, London, 1992.

Hull, C. J., '"Drows'd with the Fume of Poppies …"', Vicary Lecture, Royal College of Surgeons of England, London, 1985.

Hunt, James, 'Introductory Address on the Study of Anthropology Delivered before the Anthropological Society of London, February 24, 1863', *Anthropological Review*, Vol. 1, No. I, 1863.

Hunter, William, *On the Uncertainty of the Signs of Murder in the Case of Bastard Children*, J. Callow, Medical Bookseller, London, 1812.

Jackson, Mark, *New-born Child Murder: Women, Illegitimacy and the Courts in Eighteenth-Century England*, Manchester University Press, Manchester, 1996.

Jalland, Pat, *Australian Ways of Death: A Social and Cultural History 1840–1918*, Oxford University Press, Melbourne, 2002.

Jeffries, Stuart, 'The Naked and the Dead', *Guardian*, 19 March 2002, http://www.guardian.co.uk/print/0,3858,4376855_103680,00.html, consulted 22 November 2002.

Jones, Gareth, *Speaking for the Dead: Cadavers in Biology and Medicine*, Ashgate Dartmouth, Aldershot, England, 2000.

Jordanova, Ludmilla, 'Gender, Generation and Science: William Hunter's Obstetrical Atlas', in W. F. Bynum and Roy Porter (eds), *William Hunter and the Eighteenth-Century Medical World*, Cambridge University Press, Cambridge, 1985.

——, 'Happy Marriages and Dangerous Liaisons: Artists and Anatomy', in *The Quick and the Dead: Artists and Anatomy*, Hayward Gallery, London, 1997.

——, 'Natural Facts: A Historical Perspective on Science and Sexuality', in P. MacCormack and M. Strathern (eds), *Nature, Culture and Gender*, Cambridge University Press, Cambridge, 1980.

——, *Sexual Visions: Images of Gender in Science and Medicine between the Eighteenth and Twentieth Centuries*, University of Wisconsin Press, Madison, 1989.

Journal of Mrs. Fenton. A Narrative of her Life in India, the Isle of France (Mauritius), and Tasmania during the Years 1826–1830, with a Preface by Sir Henry Lawrence, Bart., Edward Arnold, London, 1901.

Kemp, Martin (ed.), *Dr William Hunter at the Royal Academy of Arts*, University of Glasgow Press, Glasgow, 1975.

Knox, Robert, *Great Artists and Great Anatomists; A Biographical and Philosophical Study*, John Van Voorst, London, 1852.

——, *A Manual of Artistic Anatomy, For the Use of Sculptors, Painters, and Amateurs*, Henry Renshaw, London, 1852.

——, *The Races of Man: A Fragment*, Lea & Blanchard, Philadelphia, 1850.

Kohlstedt, Sally, 'Natural Heritage: Securing Australian Materials in Nineteenth Century Museums', *Museums Australia*, December 1984.

Lake, Marilyn, 'Convict Women as Objects of Male Vision: An Historiographical Review', *Bulletin of the Centre for Tasmanian Historical Studies*, Vol. 2, No. 1, 1988.

Lang, John Dunmore, *Reminiscences of my Life and Times both in Church and State in Australia, for Upwards of Fifty Years*, Heinemann, Melbourne, 1972.

Lassek, A. M., *Human Dissection: Its Drama and Struggle*, Charles C. Thomas, Springfield, 1958.

Lawrence, Susan, 'Anatomy and Address: Creating Medical Gentlemen in Eighteenth-Century London', in Vivian Nutton and Roy Porter (eds), *The History of Medical Education in Britain*, Rodopi, Amsterdam, 1995.

——, *Charitable Knowledge: Hospital Pupils and Practitioners in Eighteenth-Century London*, Cambridge University Press, Cambridge, 1996.

Leakey, Richard and Lewin, Roger, *Origins: The Emergence and Evolution of our Species and its Possible Future*, Future, London, 1982.

Lett, Hugh, 'Anatomy at the Barber-Surgeons' Hall', Vicary Lecture, *British Journal of Surgery*, Vol. 31, 1943.

Levy, Giovanni, 'On Microhistory', in P. Burke (ed.), *New Perspectives on Historical Writing*, Polity Press, Cambridge, 1991.

Levy, M. C., *Governor George Arthur: A Colonial Benevolent Despot*, Georgian House, Melbourne, 1953.

Linebaugh, Peter, 'The Tyburn Riot against the Surgeons', in D. Hay, P. Linebaugh, J. G. Rule, E. P. Thompson and C. Winslow (eds), *Albion's Fatal Tree: Crime and Society in Eighteenth-Century England*, Allen Lane, London, 1975.

London Medical Gazette Vol. 1, No. 20, 19 April 1828.

——, Vol. 1, No. 26, 31 May 1828.

Lonsdale, Henry, *A Sketch of the Life and Writings of Robert Knox, the Anatomist 1870*, Macmillan & Co., London, 1870.

Lubbock, Sir John, *The Origin of Civilisation and the Primitive Condition of Man: Mental and Social Condition of Savages*, Longmans, Green & Co., London, 1870.

Lydekker, R., *Sir William Flower*, J. M. Dent & Co., London, 1906.

Lynn, Steven, 'Locke and Beccaria: Faculty Psychology and Capital Punishment', in William B. Thesing (ed.), *Executions and the British Experience from the 17th to the 20th Century: A Collection of Essays*, McFarland, Jefferson, NC, 1990.

Macintyre, Stuart, *A Concise History of Australia*, Cambridge University Press, Cambridge, 1999.

MacMillan, David, *Scotland and Australia 1788–1850. Emigration, Commerce and Investment*, Clarendon, Oxford, 1967.

Manne, Robert (ed.), *Whitewash: On Keith Windschuttle's Fabrication of Aboriginal History*, Black Inc. Agenda, Melbourne, 2003.

Mauss, Marcel, *The Gift: Forms and Functions of Exchange in Archaic Societies*, Cohen & West Ltd., London, 1954, translated by Ian Cunnison.

McGregor, Russell, *Imagined Destinies: Aboriginal Australians and the Doomed Race Theory 1880–1939*, Melbourne University Press, Carlton, 1997.

Meigs, J. Aitken, *Catalogue of the Human Crania in the Collection of the Academy of Natural Sciences of Philadelphia, based on the third edition of Dr Morton's 'Catalogue of Skulls'*, J. Dobson, Philadelphia, 1857.

Melville, Henry, *The History of Van Diemen's Land from the Year 1824–1835, Inclusive During the Administration to Lieutenant-Colonel George Arthur*, Horowitz Publications, Sydney, 1965.

Merwick, Donna, *Death of a Notary: Conquest and Change in Colonial New York*, Cornell University Press, Ithaca and London, 1999.

Metters, Jeremy, *The Isaacs Report: The Investigation of Events that Followed the Death of Cyril Mark Isaacs*, Department of Health, London, May 2003.

Miller, Morris E., *Pressmen and Governors: Australian Editors and Writers in Early Tasmania*, Sydney University Press, Sydney, 1974.

Mortmain, *Van Diemen's Land: A Collection of Choice Petitions, Memorials and Letters of Protest and Request from the Convict Colony of Van Diemen's Land; Written by Divers Persons, both Eminent and Lowly, and Collected and Transcribed from the Originals by Eustace FitzSymmonds, with Numerous Pages of the Manuscripts Shewn in Facsimile*, Sullivan's Cove, Hobart, 1977.

Morton, Alexander, 'Some Account of the Work and Workers of the Tasmanian Society and the Royal Society of Tasmania, from the Year 1840 to the Close of 1900', *Papers and Proceedings of the Royal Society of Tasmania*, 1901–02.

Moscucci, Ornella, *The Science of Woman: Gynaecology and Gender in England, 1800–1929*, Cambridge University Press, Cambridge, 1990.

Munro, Robin, 'Body "Art" Hits Home in Siberia', *Moscow Times*, 9 April 2001, http://www.moscowtimes.ru/stories/2001/04/09/016.html, consulted 28 November 2003.

Murray, Norman, *The Scottish Hand Loom Weavers 1790–1850: A Social History*, John Donald Publishers Ltd., Edinburgh, 1978.

Murray, Tim, 'The Childhood of William Lanne: Contact Archaeology and Aboriginality in Tasmania', *Antiquity*, Vol. 67, No. 256, 1993.

Newman, Charles, *The Evolution of Medical Education in the Nineteenth Century*, Oxford University Press, London, 1957.

Nicholas, Stephen (ed.), *Convict Workers: Reinterpreting Australia's Past*, Cambridge University Press, Cambridge, 1988.

Nicholls, Mary (ed.), *The Diary of the Reverend Robert Knopwood 1803–1838 First Chaplain of Van Diemen's Land*, Tasmanian Historical Research Association, Hobart, 1977.

Oxley, Deborah, *Convict Maids: The Forced Migration of Women to Australia*, Cambridge University Press, Cambridge, 1996.

Paffen, Paul, 'The Bushrangers of Van Diemen's Land', in Tim Bonyhady and Andrew Sayers (eds), *Heads of the People: A Portrait of Colonial Australia*, National Portrait Gallery, Canberra, 2000.

——, 'Forgotten Faces? Portraits and Other Images of the Convict in Van Diemen's Land', *Tasmanian Historical Research Association, Papers and Proceedings*, Vol. 46, No. 2, June 1999.

Parliamentary Papers, Tasmania, Legislative Council, General Hospital, Hobart Town, Report of Commission, January 1877.

——, Report of Select Committee, September 1875.

Parrott, Jennifer, 'Elizabeth Fry and Female Transportation', *Tasmanian Historical Research Association, Papers and Proceedings*, Vol. 43, No. 4, 1996.

Payne, H. S., 'A Statistical Study of Female Convicts in Tasmania, 1843–53', *Papers and Proceedings of the Tasmanian Historical Research Association*, Vol. 9, No. 2, 1961.

Perkin, H., 'The Old Society', in H. Perkin (ed.), *The Origins of Modern English Society 1780–1880*, Routledge, London, 1969.

Persaud, T. V. N., *A History of Anatomy: The Post-Vesalian Era*, Charles C. Thomas, Springfield, 1997.

Petropoulos, Thrasy, 'Seat at the Autopsy Sideshow', *BBC News*, http://news.bbc.co.uk/1/hi/health/2497889.stm, consulted 22 November 2002.

Petrow, Stefan, 'The Last Man: The Mutilation of William Lanne in 1869 and its Aftermath', *Australian Cultural History*, No. 16, 1998.

Piesse, E. L., 'The Foundation and Early Work of the Society; with some Account of Earlier Institutions and Societies in Tasmania', *Papers and Proceedings of the Royal Society of Tasmania*, 1913.

Plomley, N. J. B., *Friendly Mission: The Tasmanian Journals and Papers of George Augustus Robinson, 1829–1834*, Tasmanian Historical Research Association, Hobart, 1966.

——, 'A List of Tasmanian Aboriginal Material in Collections in Europe', *Records of the Queen Victoria Museum*, New Series, No. 15, 1962.

——, 'Thomas Bock's Portraits of the Tasmanian Aborigines', *Records of the Queen Victoria Museum*, New Series, No. 18, 1965.

——, *Weep in Silence: A History of the Flinders Island Aboriginal Settlement; with the Flinders Island Journal of George Augustus Robinson, 1835–1839*, Blubber Head Press, Hobart, 1987.

Porter, Roy, 'History of the Body', in Peter Burke (ed.), *New Perspectives on Historical Writing*, Polity Press, Cambridge, 1991.

——, 'Hospitals and Surgery', in Roy Porter (ed.), *Cambridge Illustrated History of Medicine*, Cambridge University Press, Cambridge, 1996.

'Public Autopsy Stirs Debate in London', *Japan News Today*, 21 November 2002, http://www.japantoday.com/gidx/news239513.html, consulted 29 March 2004.

Pybus, Cassandra, *Community of Thieves*, William Heinemann, Port Melbourne, 1991.

Radzinowski, Leon, *A History of English Criminal Law and its Administration from 1750. Volume 1: The Movement for Reform*, Stevens & Sons Ltd, London, 1948.

Rae, Isobel, *Knox the Anatomist*, Oliver & Boyd, Edinburgh and London, 1964.

Rae-Ellis, Vivienne, *Black Robinson: Protector of Aborigines*, Melbourne University Press, Carlton, 1988.

——, *Trucanini: Queen or Traitor?*, Australian Institute of Aboriginal Studies, Canberra, 1981, second edition.

Reifler, Douglas, '"I actually don't mind the bone saw": Narratives of Gross Anatomy', *Literature and Medicine*, Vol. 15, No. 2, 1996.

Reilly, Dean, 'Interview Professor Gunther von Hagens', June 2002, http://www.artsnet.org.uk/pages/gunthervonhagens2.html, consulted 29 March 2004.

Reynolds, Henry, *Fate of a Free People*, Penguin, Ringwood, 1995.

Richards, Evelleen, 'The "Moral Anatomy" of Robert Knox: The Interplay between Biological and Social Thought in Victorian Scientific Naturalism', *Journal of the History of Biology*, Vol. 22, No. 3, 1989.

Richardson, Ruth, *Death, Dissection and Destitution*, Penguin, Harmondsworth, 1988, 2001.

Rimmer, W. G., *Portrait of a Hospital: The Royal Hobart*, Royal Hobart Hospital, Hobart, 1981.

Robinson, Portia, *The Women of Botany Bay*, Penguin, Ringwood, 1993, first published 1988.

Robson, Lloyd, *A History of Tasmania: Volume I: Van Diemen's Land from the Earliest Times to 1855*, Oxford University Press, Melbourne, 1983.

——, *A History of Tasmania: Volume II: Colony and State from 1856 to the 1980s*, Oxford University Press, Melbourne, 1991.

Roe, Michael (ed.), *The Flow of Culture: Tasmanian Studies*, Australian Academy of the Humanities, Canberra, 1988.

Romains, Jules, *The Death of a Nobody*, Alfred A. Knopf, New York, 1944, translated by Desmond MacCarthy and Sidney Waterlow, second edition.

Ross, James, 'An Essay on Prison Discipline in which is Detailed the System Pursued in Van Diemen's Land', *The Van Diemen's Land Annual and Hobart-town Almanack for the Year 1833*, James Ross, Hobart, 1834.

Roth, H. Ling, *The Aborigines of Tasmania*, Fullers Bookshop Pty Ltd, Hobart, first published 1890, facsimile of second edition, 1968.

Roth, Richard, 'London's Cadaver: a Reality Show Low?', *CBS News*, 20 November 2002, http://www.cbsnews.com/stories/2002/11/20/world/main530223.shtml, consulted 29 March 2004.

Roughhead, William (ed.), *Burke and Hare*, William Hodge & Company Limited, London, 1948.

Royal Liverpool Children's Inquiry: Report, ordered by The House of Commons to be printed January 30, 2001.

Royal Society of Tasmania, 'Some Account of the Work and Workers of the Tasmanian Society and the Royal Society of Tasmania', *Papers and Proceedings, Royal Society of Tasmania*, 1900–01.

Ryan, Lyndall, *The Aboriginal Tasmanians*, University of Queensland Press, St Lucia, 1981.

——, 'The Governed: Convict Women in Tasmania 1803–1853', *Bulletin of the Centre for Tasmanian Historical Studies*, Vol. 3, No. 1, 1990–91.

Sappol, Michael, *A Traffic of Dead Bodies: Anatomy and Embodied Social Identity in Nineteenth-Century America*, Princeton University Press, Princeton, 2002.

Savery, Henry, *The Hermit in Van Diemen's Land*, University of Queensland Press, St Lucia, 1964, first published 1829.

Sawday, Jonathon, *The Body Emblazoned: Dissection and the Human Body in Renaissance Culture*, Routledge, London, 1995.

Schiebinger, Londa, *The Mind has no Sex? Women in the Origins of Modern Science*, Harvard University Press, Cambridge, Mass., 1989.

Seager, Philip S., *Hobart General Hospital Centenary Celebration: Epitome of its History*, John Vail, Government Printer, Hobart, 1921.

Sharpe, J. A., *Judicial Punishment in England*, Faber & Faber, London, 1990.

Shaw, A. G. L., *Sir George Arthur, Bart. 1784–1854*, Melbourne University Press, Carlton, 1980.

Shelton, D. C. (ed.), *The Parramore Letters from Van Diemen's Land*, privately published, Hobart, 1993.

Smedley, Bunny, 'Habeas Corpus: The Old Taboos are the Best Ones', *Electric Review: Politics, Art & Literature*, 28 November 2002, http://www.electric-review.com/archives/000182.html, consulted 29 March 2004.

Smith, Thomas Southwood, 'A Lecture Delivered over the Remains of Jeremy Bentham, Esquire, in the Webb-Street School of Anatomy and Medicine on the 9th of June, 1832', Effingham Wilson, London, 1832.

——, *Use of the Dead to the Living*, Printed for Subscribers on the *Lancet* Press, London, 1827.

Somerville, J., 'The Royal Society of Tasmania, 1843–1943', *Papers and Proceedings of the Royal Society of Tasmania*, December 1944.

South, John Flint, *Memorials of John Flint South, Twice President of Royal College of Surgeons and Surgeon to St Thomas's 1841–63*, collected by the Rev. Charles Lett Feltoe, John Murray, London, 1884.

——, *The Dissector's Manual. A New Edition, with Additions and Alterations*, J. Mawman, London, 1825.

Sprod, Dan, *Alexander Pearce of Macquarie Harbour: Convict Bushranger Cannibal*, Cat and Fiddle Press, Hobart, 1977.

Stepan, Nancy, *The Idea of Race in Science*, Macmillan in association with St Antony's College, Oxford, 1982.

Stocking, George W. Jr, *Race, Culture, and Evolution: Essays in the History of Anthropology*, The Free Press, New York, 1968.

——, *Victorian Anthropologists*, The Free Press, New York, 1987.

——, 'What's in a Name? The Origins of the Royal Anthropological Institute, 1837–71', *Man*, Vol. 6, No. 3, 1971.

Tardif, Phillip, *Notorious Strumpets and Dangerous Girls: Convict Women in Van Diemen's Land 1803–1829*, Angus & Robertson, North Ryde, NSW, 1990.

Tasmanian Historical Research Association, *Van Diemen's Land: Copies of all Correspondence Between Lt Gov Arthur and His Majesty's Secretary of State for the Colonies, on the Subject of the Military Operations Lately Carried on against the Aboriginal Inhabitants of Van Diemen's Land*, Tasmanian Historical Research Association, Hobart, 1971.

Taylor, Shepherd T., *The Diary of a Medical Student During the Mid-Victorian Period 1860–1864*, Jarrold & Sons, Ltd, Norwich, 1927.

Thesing, William B. (ed.), *Executions and the British Experience from the Seventeenth to the Twentieth Century: A Collection of Essays*, McFarland & Company, Jefferson, North Carolina and London, 1990.

Thompson, E. P., *The Making of the English Working Class*, Penguin, Harmondsworth, 1986.

Timbs, John, *Curiosities of London, Exhibiting the Most Rare and Unusual Objects of Interest in the Metropolis; with Nearly Fifty Years' Personal Recollections*, David Bogue, London, 1855.

Trollope, Anthony, *Australia and New Zealand*, Chapman & Hall, London, 1876, Vol. 2, third edition.

Turnbull, Clive, *Black War: The Extermination of the Tasmanian Aborigines*, Cheshire-Lansdowne, Melbourne, 1965, first published 1948.

Turnbull, Paul, 'Ancestors, not Specimens: Reflections on the Controversy over the Remains of Aboriginal People in European Scientific Collections', in Ken Riddiford, Eric Wilson and Barry Wright (eds), *Contemporary Issues in Aboriginal and Torres Strait Islander Studies: 4. Proceedings of the 4th National Conference*, Cairns College of Technical and Further Education, Cairns, 1993.

——, 'Enlightenment Anthropology and the Ancestral Remains of Australian Aboriginal People', in Alex Calder, Jonathon Lamb and Bridget Orr (eds), *Voyages and Beaches: Pacific Encounters, 1769–1840*, University of Hawaii, Honolulu, 1999.

——, '"Ramsay's Regime": The Australian Museum and the Procurement of Aboriginal Bodies c. 1874–1900', *Aboriginal History*, No. 15, 1991.

——, 'Science, National Identity and Aboriginal Body Snatching in Nineteenth Century Australia', *Working Papers in Australian Studies*, No. 65, Sir Robert Menzies Centre for Australian Studies, University of London, 1991.

——, 'To What Strange Uses: The Procurement and Use of Aboriginal People's Bodies in Colonial Australia', *Voices: the Quarterly Journal of the National Library of Australia*, 1994.

Tylor, Edward B., 'On the *Tasmanians* as Representatives of *Paolaeolithic Man*', *Journal of the Anthropological Institute*, Vol. XXIII, 1893.

'UCLA Illegally Sold Human Body Parts', 8 March 2004, KRON News, http://www.kron.com/Global/story.asp?s=1696561, consulted 16 March 2004.

Walker, Alan and Shipman, Pat, *The Wisdom of Bones: In Search of Human Origins*, Weidenfeld & Nicolson, London, 1996.

Walker, Bret, *Inquiry into Matters Arising from the Post-mortem and Anatomical Examination Practices of the Institute of Forensic Medicine: Report*, Government of New South Wales, 2001.

Warner, John Harley, *Against the Spirit of System: The French Impulse in Nineteenth-Century American Medicine*, Princeton University Press, Princeton, 1998.

West, John, *The History of Tasmania with Copious Information Respecting the Colonies of New South Wales, Victoria, South Australia, &c., &c., &c.*, Angus & Robertson Publishers, London, 1852.

Windschuttle, Keith, *The Fabrication of Aboriginal History, Volume One, Van Diemen's Land 1803–1847*, Macleay Press, Sydney, 2002.

Winter, Gillian (ed.), *Tasmanian Insights: Essays in Honour of Geoffrey Thomas Stilwell*, State Library of Tasmania, Hobart, 1992.

Wolfe, Patrick, *Settler Colonialism and the Transformation of Anthropology: The Politics and Poetics of an Ethnographic Event*, Cassell, London, 1999.

Wyman, Jeffries, 'Observations on the Skeleton of a Hottentot', *Anthropological Review*, Vol. III, No. VIII, 1865.

Yarwood, A. T. and Knowling, M. J., *Race Relations in Australia: A History*, Methuen Australia, North Ryde, NSW, 1982.

THESES

MacDonald, Helen, 'Human Remains: Episodes in Nineteenth-Century Colonial Human Dissection', PhD Thesis, University of Melbourne, 2002.

Winter, Gillian, '"For … the advancement of science": The Royal Society of Tasmania 1843–1885', BA (Hons) Thesis, University of Tasmania, 1972.

INDEX

Page references in **bold** indicate illustrations.